# Binary Junction Transistors
## From Scratch
### Richard (Dick) Whipple

Richard (Dick) Whipple
Visit my website at www.whippleway.com

Printed in the United States of America

First Printing
Amazon Paperback

ISBN 9798870402598

# Contents

# Introduction

As with all his "From Scratch" books, the author assumes that the reader has no prior knowledge of the book's subject, which, in the present case, is the *binary junction transistor* or *BJT*. As always, we start with basic ideas and build on them until we are ready to apply our knowledge to an actual application.

Specifically, we will apply BJT theory and principles of circuit design to build a BJT high-fidelity amplifier with these specifications:

1. Input sensitivity - 1-volt peak-to-peak sinusoid of frequency 1000 Hz to produce at least 5 watts output in an 8 Ω speaker load.
2. Input load - No less than 100 KΩ.
3. Frequency response – At least ±1 DB from 20 Hz to 20,000 Hz.
4. Harmonic distortion - At 1000 Hz, less than 1%.

Along the way, we will verify our theory using software to simulate the circuit. Simulations provide a convenient way for us to try design ideas before committing them to actual circuits. The simulation also allows us to optimize circuit values by rerunning simulations and choosing the component value that works best. When we are satisfied with the simulated model, we will *breadboard* an actual circuit and verify that it meets its specifications.

Simulations that model BJTs and other components use a software language called SPICE (Simulation Program with Integrated Circuit Emphasis). While a variety of software companies implement SPICE models, the author chose *MultisimLive* (https://www.multisim.com/). It requires a small subscription fee, but it is well worth the cost. A second source is *Circuit Lab* (https://www.circuitlab.com/). It is free and works quite well.

Simulations are dependable, but, like all models, they have limits and are not perfect. That is why we follow up by breadboarding the design. Besides verifying our simulation, breadboarding also permits us to discover aspects of the design not covered by the simulation model. For example, our simulation might not take component heating into account, making the breadboard step essential before fully validating the design.

Because models, like SPICE, play such a vital role in all that we do, we devote the next chapter to considering the concept of *models* and, most significantly, their limitations.

# Chapter 1 – Models

The purpose of a *model* is to help us understand how a physical object or process works. It need not be an exact representation but one that conveys a level of understanding consistent with the needs of a target audience. Thus, the model used by a scientist or engineer is apt to be quite different from one directed at a layperson. The former might provide mathematically precise insight, while the latter may only convey a basic conceptual idea.

For our "From Scratch" exploration of BJTs, we will use models higher than the layperson level but not at the science/engineering level. Sometimes, models will consist of images that help us visualize how BJTs work. Other models will use mathematical formulas that pin down the operational characteristics of BJTs. Eventually, we will utilize SPICE models in the design phase of our BJT circuits.

One thing to keep in mind, however, is that models serve a specific purpose, and we must not extend them beyond their intended purpose. Nor should we infer a characteristic beyond that specified for the model. Take, for example, the fact that SPICE modeling may not take the thermal properties of BJTs into account. We may produce a perfect circuit that, when breadboarded, self-destructs with thermal runaway!

Another example of model limitations is the *modified Bohr model* of the lithium atom below.

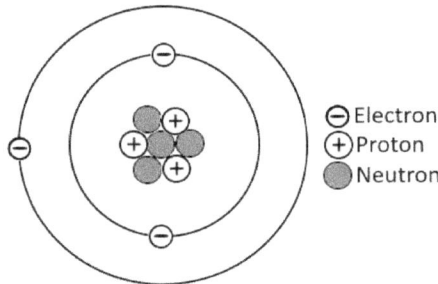

⊖ Electron
⊕ Proton
⬤ Neutron

For this book, the modified Bohr model consists of a compact, positively charged nucleus with a fixed number of protons and neutrons. In the case of lithium, the nucleus contains three protons and four neutrons. The circles surrounding the nucleus represent *shells* that contain negatively charged electrons. Electrons in each shell possess a fixed energy, with the lowest energy electrons in the first shell. Since the number of positively charged protons is the same as the number of negatively charged electrons, the electrical charge of the atom equals zero.

The modified Bohr model helps illustrate how atoms combine to make up molecules or crystals. But that is where we draw the line. For instance, we cannot infer the relative sizes of the atomic particles from the model. Nor should we assume that electrons "orbit" the nucleus like minor planets in a solar system. There are more sophisticated models, such as the *quantum model*, that more accurately represent the true nature of atoms.

We classify the modified Bohr model as a *qualitative model* because it does not provide mathematical representations. Later, we will encounter the BJT *hybrid model*, a *quantitative model* that includes parameter values and their mathematical relationships. With it, we will study the behavior of BJTs in electrical circuits. Here again, the hybrid model has limitations. The electrical behavior it simulates is never exact, nor does it apply to all situations. **To reiterate, models serve a specific and limited purpose, and we must not extend them beyond their intended purpose.**

We begin in the next chapter with basic BJT theory. Specifically, we will use the modified Bohr model and quantum theory to explain electrical conduction in solids.

# Chapter 2 – Electrical Conduction in a Solid

Scientists classify solid materials as to their electrical conduction as *insulators*, *semiconductors*, and *conductors*. The classification is based on their capacity to support electron flow in the presence of an electric field. We call this capacity *conductivity* and this electron flow *electrical current*. From a qualitative point of view, insulators have no conductivity, semiconductors have limited conductivity, and conductors have large conductivity.

To explain the reason for differences in conductivity, we need to consider the atomic structure of the material. To simplify our approach, we will focus on solids existing in crystalline form and use our modified Bohr model to show this form.

From quantum theory, we learn that electrons in the modified Bohr model can only occupy specific energy levels. This is the case for an isolated atom. When in solids, however, many atoms exist in proximity to each other, and quantum effects cause the fixed energy levels to broaden out to *energy bands*. See the figure below.

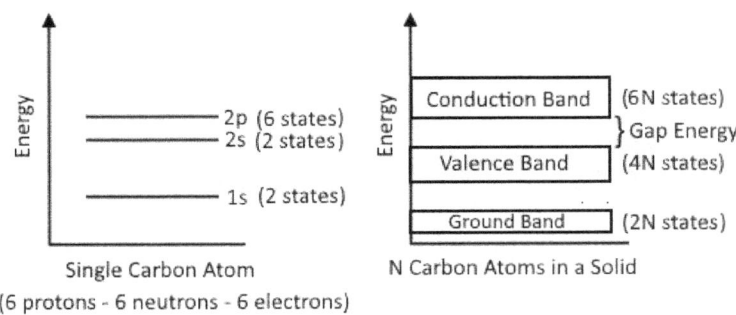

Single Carbon Atom
(6 protons - 6 neutrons - 6 electrons)

N Carbon Atoms in a Solid

The image on the left shows the energy levels for a single carbon atom. An energy level contains a specific number of states filled with electrons. Quantum effects take hold as we pack large numbers of carbon atoms (say N) into a solid mass, as shown in the right image. Instead of electrons taking on a single energy state, they can assume any energy within a fixed band of energy.

For large values of N, the carbon atom bands take on these characteristic behaviors. The lowest energy band is very stable, and electrons there stay put! The *valence band* is also stable, but a movement to the *conduction band* is possible if, with the addition of heat energy, valence band electrons acquire the *gap energy*. After jumping to the *conduction band*, electrons gain the mobility needed to move between atoms under the influence of an electric field. Electron flow consists of the movement of electrons with sufficient energy to be in the conduction band.

For most solids, we only need to consider the valence and conduction bands, as shown below.

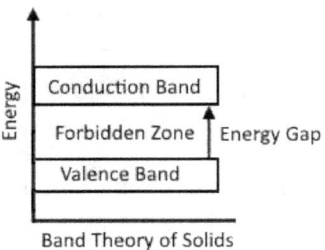

Band Theory of Solids

Because electrons cannot reside in the energy gap, we call this region the *forbidden zone*. Further, the size of the energy gap determines whether the solid is an insulator, semiconductor, or conductor. See the figure below.

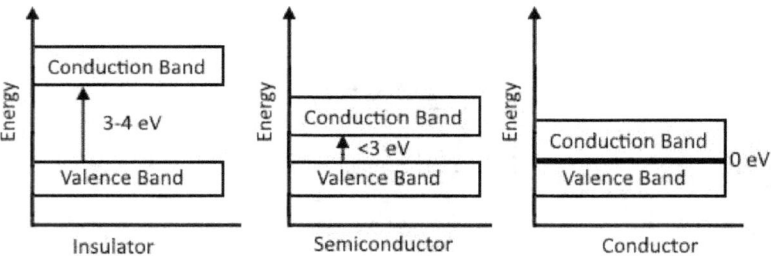

For insulators at room temperature, the energy gap is large enough that no valence electrons reach the conduction band so that no electrical current can flow. Conductors, on the other hand, have no energy gap, so electrons readily inhabit the conduction band, permitting sizeable current flow. With semiconductors, the energy gap is smaller than that of insulators, allowing a few electrons to find their way into the conduction band. Semiconductor current flow at room temperature, therefore, is slight compared to that of conductors.

Glass, ceramic, and plastic are examples of insulators. Good conductors include copper, aluminum, and silver. Of the possible semiconductor materials, silicon and germanium are the most desirable for our purposes.

In the next chapter, we discuss how to control silicon and germanium conductivity as a prelude to creating a *semiconductor junction*.

# Chapter 3 – Increasing Semiconductor Conductivity

As noted in Chapter 2, silicon and germanium are semiconductors because of their small energy gap, 1.12 eV and 0.75 eV, respectively. Both materials can exist in pure crystal form, which we refer to as *intrinsic silicon* and *intrinsic germanium*. In crystal form, these atoms arrange themselves in a way that creates *covalent bonds* or *shared electron bonds,* as shown below.

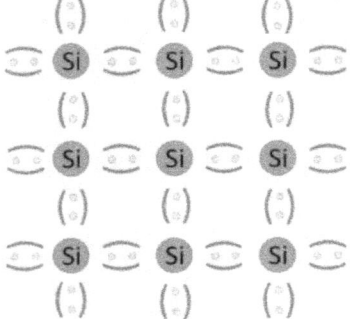

The four valence electrons of each atom pair up with the valence atoms of adjacent atoms. At room temperature, a few outer electrons transition to the conduction band, giving silicon its semiconductor designation.

We can increase the number of valence electrons that jump to the conduction band by heating the semiconductor. We can increase the conductivity without heating by mixing in atoms with five valence electrons. The "mixing" we refer to as *doping*. We refer to such a doped mixture as *extrinsic silicon* and *extrinsic germanium*. We call the added atom the *donor* and the semiconductor atom the *acceptor*. Donor substances include phosphorus, arsenic, antimony, and bismuth. Shown below is the silicon crystal *doped* with arsenic.

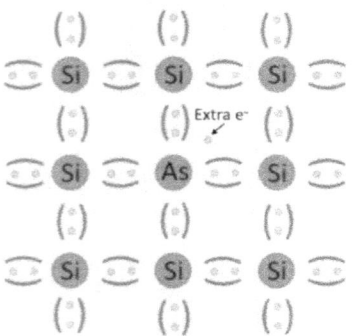

Four of the arsenic valence electrons establish covalent bonds with adjacent silicon atoms. At room temperature, the fifth arsenic valence electron is free to move into the conduction band. Since the current flow thus created is due primarily to negatively charged electrons moving in the conduction band, we call this doped substance an *N-type semiconductor*.

We can create a different type of charge carrier if we dope intrinsic silicon with boron, aluminum, or gallium. These atoms have only three valence electrons. When we dope silicon with one of these, we create an unfilled valence position or *hole* in the crystalline structure. See in the figure below what happens when we dope silicon with boron.

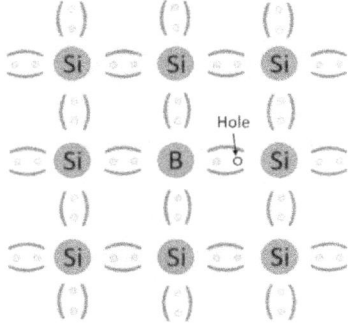

The energy required to fill the hole is just slightly above the valence band energy, so it is easy for an electron from the valence band of a neighboring atom to jump into it. For this reason, we call the boron the *acceptor atom,* and silicon becomes the *donor atom*.

As noted, a nearby valence electron can jump into this hole, leaving a hole in the donor silicon atom. This new hole creates a positive acceptor atom, ready to receive another donor electron. This movement of electrons from hole to hole within the valence band constitutes electron flow different from that in the conduction band. To distinguish it from electron flow in the conduction band, we think of it as holes flowing in the opposite direction in the valence band. Since this current flow is due primarily to positively charged holes moving in the valence band, we call this doped substance a P-*type semiconductor*.

Note: When discussing P-type semiconductors, we refer to holes as moving when they do not move but merely appear to move as electrons jump from one hole to another. We accept this misconception because it distinguishes the differing behavior of P-type and N-type semiconductors. **The fact is that only electrons move!**

As we have seen, conduction in semiconductors occurs in two ways. Valence band electrons provide most of N-type semiconductor conduction. Thus, electrons are *the majority charge carriers* in N-type semiconductors. In addition, electrons traversing the energy gap into the conduction band leave holes in the valence band that low-energy electrons can jump between, producing a small amount of hole current flow. Thus, in N-type semiconductors, we say holes are *minority charge carriers*.

Holes are the majority charge carriers in P-type semiconductors since they are responsible for most of the conduction. Electrons in the conduction band provide a small conduction capability. Thus, in P-type semiconductors, we say electrons are minority carriers.

An interesting point is that electron flow in N-type semiconductors (called *mobility*) is more unrestricted than hole flow in P-type semiconductors. This makes N-type devices both more efficient and faster. As a result, they are more frequently employed in high-frequency and high-power applications.

N-type and P-type semiconductor conduction is not particularly useful. It is how we employ them in combination that leads to the creation of devices that serve a wide variety of electronic applications.

In the chapters that follow, we explore these semiconductor devices along with their applications.

# Chapter 4 – The P-N Junction Diode

As suggested in the previous chapter, combining P-type and N-type semiconductor materials leads to the creation of a wide range of practical electronic devices. In this chapter, the device we will focus on is the *semiconductor diode*. A diode is a two-terminal device that conducts electrical current in only one direction. An *ideal diode* presents zero resistance in the forward direction and infinite resistance in the reverse direction. As we shall see, the semiconductor diode comes close to exhibiting these characteristics and thus finds a broad range of applications in electronic applications.

Consider a crystal of silicon doped in such a way that one side is P-type, and the other side is N-type. See the figure below.

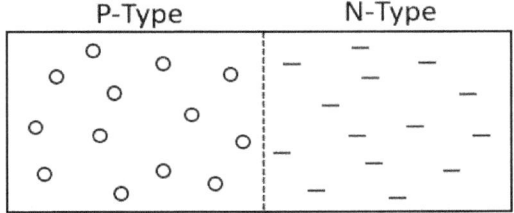

Only shown are the majority carriers, holes in the P-type, and electrons in the N-type. Initially, the presence of equal numbers of opposite charges in each half leaves the crystal electrically neutral in charge.

At room temperature, the majority carriers on both sides of the *junction* (the dotted line on the figure) are free to move within the crystal. The majority carriers (electrons and holes) begin to defuse across the junction, where they neutralize one another, creating a *depletion* or *space charge zone*. See the figure below.

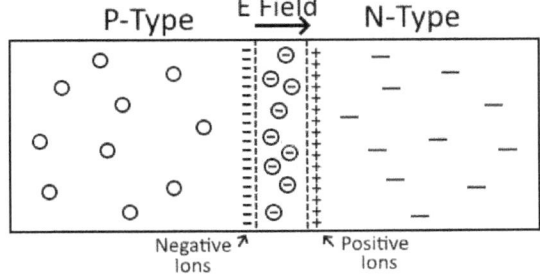

Removing holes from the P-type material leaves negative ions adjacent to the P-edge of the depletion zone. Similarly, removing electrons from the N-type leaves positive ions adjacent to the N-edge of the depletion zone. The differing ionic charge across the depletion zone produces an electrical field that opposes the movement of both holes and electrons into the depletion zone. Diffusion continues until the electrical field potential stops further migration of holes and electrons. At room temperature in silicon, this *electric field potential* is about 0.7 volts; for germanium, it is about 0.3 volts.

In this state, the depletion zone lacks charge carriers, so the semiconductor acts like an insulator. But, unlike a true insulator, the semiconductor's conductive behavior varies with an externally applied voltage. To see this, consider the circuit arrangement below.

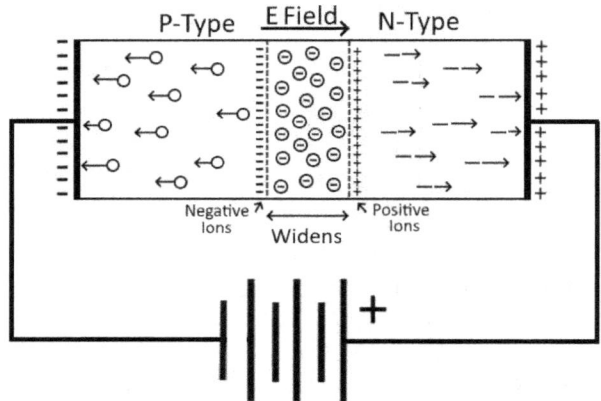

Suppose we affix metal plates with attached wires to the ends of the semiconductor and apply battery voltage with the polarity shown. The negative charge on the P plate attracts holes in the P-type material and moves them away from the depletion zone. Similarly, a positive charge on the N plate attracts electrons away from the depletion zone. By removing the majority carriers from the junction, we widen the depletion zone and increase the internal electrical field. The remaining majority carriers do not have sufficient energy to overcome the field potential, further reducing conduction and current flow. Under this external voltage arrangement, we say that the P-N junction is *reverse-biased*.

In the circuit arrangement below, we reverse the battery polarity.

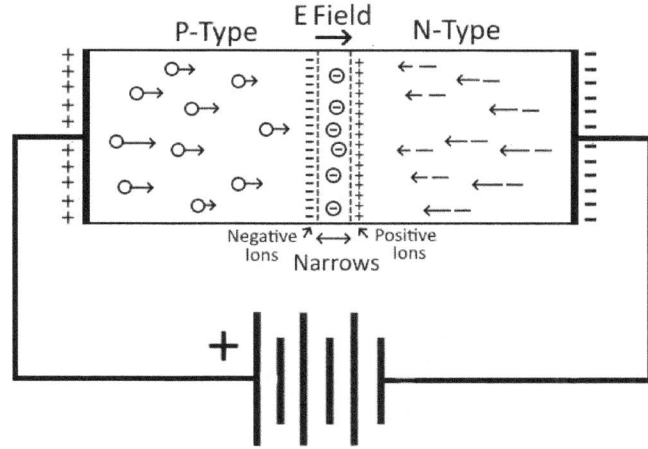

The charge on the plates now moves the majority carriers away from the end plates, flooding the depletion zone and shrinking it. The reduced electric field allows the majority carriers to cross the junction. Electrical current increases slowly with external voltage until it reaches the depletion potential (0.7 volts for silicon or 0.3 volts for germanium). At this point, the current rises sharply, and the semiconductor behaves more like a conductor. We say that, under this condition, the P-N junction is *forward-biased*.

To summarize, we have created a semiconductor device that has low conductivity when reverse-biased and high conductivity when forward-biased. If we encapsulate the device in an insulating material, the result is a semiconductor diode. When used with small signals, such as a radio detector, the encapsulating material is often glass. For rectifier use where a high current generates excessive heat, manufacturers use an encapsulating polymeric material that tolerates elevated temperatures and exhibits good heat conductivity.

The figure below shows typical characteristics of both germanium and silicon diodes.

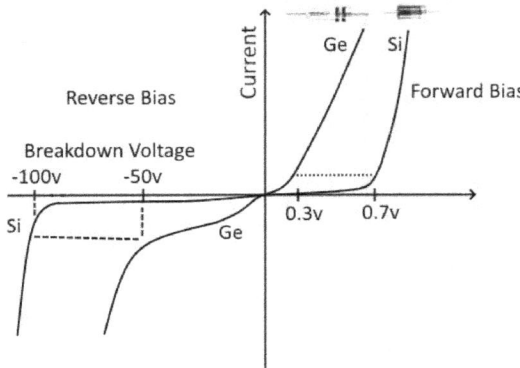

The diode's behavior when forward bias is as we described above. Of course, at some forward current, the junction will overheat and, in time, destroy the diode. For this reason, manufacturers rate diodes by maximum current.

In reverse bias, note the presence of a small current caused by minority carriers. Recall that holes are the minority charge carriers in N-type material. A small number of positively charged holes near the junction have enough energy to pierce the energy gap (described as "tunneling"). When they do, the electric potential (positive to negative) sweeps them across the depletion zone and into the P-type material. We call this reverse biased flow of holes *leakage current*. In the figure above, the leakage current is greater for germanium than silicon diodes.

As the reverse voltage increases, the electric field across the depletion zone increases. Eventually, two effects cause a sudden increase in current. One is the *avalanche effect* caused by the breakdown of the covalent bonds in the crystal and the unleashing of a flood of electrons. The other is the *Zener effect* caused by an excessive field potential across the depletion zone. The result is a sudden, large flow of current ending low conduction and ceasing diode behavior.

The knee of the reverse bias curve occurs at the diode's *breakdown voltage*. The breakdown voltage for germanium diodes is usually 100 volts or less. Silicon diodes, on the other hand, are available with breakdown voltages in thousands of volts!

When we use silicon diodes in rectifier applications, specifications include the current carrying capacity <u>and</u> the breakdown voltage, more commonly specified as the *peak-inverse voltage* or *PIV*. A rectifier's PIV indicates the maximum reverse voltage that safely avoids breakdown.

Because of the low forward turn-on voltage, germanium diodes find application in low-level signal applications. For instance, AM and FM radios use germanium diodes as detectors. While there are specialty germanium diodes made with P-N junctions like that described above, the *point contact* type is more prevalent. See the link below for a description of point contact germanium diode construction.

Link: https://www.eeeguide.com/point-contact-diode-construction-and-working/

To summarize, an ideal diode has zero forward resistance, infinite reverse resistance, and no breakdown voltage. Silicon diodes come closest. When forward biased, their internal resistance drops to as low as 0.1 ohm with a current carrying capacity in the high ampere range. In reverse bias, their leakage current is low, and their breakdown voltage exceeds most practical requirements.

As we have seen, electron current flows from the N-type to P-type material. *Convention current* flows in the opposite direction, i.e., from P-type to N-type. Standard cylindrical packaging of diodes indicates conventional current flow direction with a dark band on the N end of the diode. The standard schematic symbol for a diode also uses conventional current notation. As shown below, it consists of a bar following an arrow pointing toward conventional current flow.

Conventional Current Flow

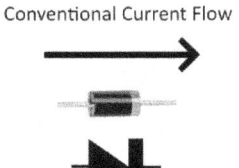

Note: In the early days of electrical experimentation, people thought that positive charges flowed from the positive battery terminal to the negative battery terminal. This gave rise to symbolism that showed the current flow in that direction. Later, scientists discovered that electron flow was in the opposite direction. By this time, positive charge flow was so widespread and ingrained that it acquired the name *conventional current flow* and continued in use. For circuit analysis, the direction does not matter if we use a consistent sign convention in setting up the solution.

In addition to rectifiers and small signal use, other applications of P-N junctions developed over time. For example, the Zener effect gave rise to specialized *Zener diodes* used for voltage regulation. Tunnel diodes take advantage of the tunneling effect to create high-frequency oscillators and amplifiers. Searching "types of diodes" online will yield internet sites with information about diode types and their applications.

In the next chapter, we see how adding a second junction creates the binary junction transistor.

# Chapter 5 – The Binary Junction Transistor

John Bardeen, Walter H. Brattain, and William B. Shockley of the Bell Labs invented the *Bipolar Junction Transistor* or *BJT* in 1947. As the name implies, it consists of two junctions, as shown below.

We designate this as an *NPN BJT* because we sandwich a P-type semiconductor between two N-type semiconductors. Note that the P-type is much thinner than the N-type. As we will see shortly, this will play a key role in BJT's operation.

We can also make a *PNP BJT* by sandwiching a thin N-type semiconductor between two P-types. See the figure below.

Both NPN and PNP BJTs behave similarly in electronic circuits except for the reversed polarity of external voltages. Internally, they differ in that electrons play a dominant conduction role in NPN BJTs, while holes are dominant in PNP BJTs. Since electrons move more rapidly in the conduction band than holes in the valence band, NPN BJTs work better in electronic circuits involving high frequencies or fast switching time.

The BJT acts as a current amplifier. To see how, consider the circuit below.

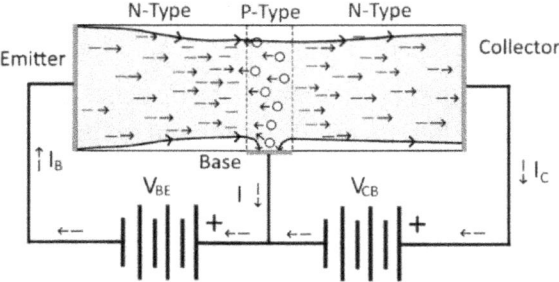

We attach metal plates to the N-type ends and the P-type edge. The N-type ends are the *emitter* and *collector*. Electrons "emit" from the emitter and "collect" at the collector, thus giving rise to the names. The P-type edge is the *base*. The term "base" has a more obscure origin. One suggested source comes from the first BJT developed at Bell Labs. The BJT's "base" was a metal plate on which the other semiconductor components rested.

Referring to the circuit above, voltage $V_{BE}$ forward-biases the base-emitter junction, while voltage $V_{CB}$ reverse-biases the collector-base junction. Under this biasing, we might expect the emitter-base current to be greater than zero ($I_E > 0$) and the collector-base current to be zero ($I_C = 0$). This would be the case if the P-type were thick, but as noted above, it is very thin. The forward bias base-emitter junction draws electrons, as we would expect. Because of the P-type's thinness, electrons sweep across it and into the collector N-type. Once they are there, the electric potential sweeps them to the collector plate. The result is a collector current that, except for a small base current, is equal to the emitter current.

We can represent this situation mathematically by noting that:

$I_C = I_E - I_B$. (Eqn. 5-1)

$I_B << I_E$ (<< meaning "much, much less"), we further can write,

$I_C = \alpha \cdot I_E$ (Eqn. 5-2)

where typically $0.95 < \alpha < 1$ (meaning $\alpha$ ranges from 0.95 to 1).

Alternatively, we can represent the collector current $I_C$ in terms of $I_B$ using the equation:

$I_C = \beta \cdot I_B$ (Eqn. 5-3)

$\beta$ (beta) typically ranges from 20 to 300. Since $I_C = \beta \cdot I_B$ and $\beta > 1$, we confirm the premise that a BJT is a *current amplifier*.

Note: $\beta$ is the *static* or *DC current gain*. Later, we will find that BJT is also capable of AC current gain, which we will designate with a different symbol.

We can relate $\alpha$ and $\beta$ by first substituting Eqns. 5-2 and 5-3 in Eqn. 5-1.

$$\beta \cdot I_B = \frac{\beta \cdot I_B}{\alpha} + I_B$$

Dividing by $I_B$ and rearranging, we find that,

$$\alpha = \frac{\beta}{\beta + 1} \quad \text{(Eqn. 5-4)}$$

Most often, the manufacturer's BJT data sheets specify $\beta$ rather than $\alpha$. When needed, we can calculate $\alpha$ using Eqn. 5-4.

Note: We sometimes see $\beta$ written as $h_{FE}$ where the capital "FE" indicates a DC value. Later, we will use non-capitals to represent AC (small signal) values. Then, we will use $h_{fe}$ to represent AC (small signal) current gain. Typically, $h_{fe}$ is larger than $h_{FE}$.

We could offer a similar explanation for how a PNP BJT works. The only differences are (1) the polarity of the voltage supply is reversed, and (2) the holes are the majority charge carriers.

The figure below shows the standard schematic symbol for the NPN and PNP BJT.

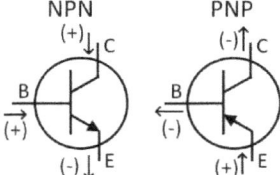

In both cases, the emitter arrows show the direction of conventional current flow.

The arrangement of components below is known as a *common emitter (CE) BJT circuit*. The word "common" in this context means the "E" (the emitter) in "CE" connects to the circuit's power common (the negative supply for NPN as shown below or the positive supply for PNP). In addition to CE, we have the *common collector* (CC) configuration with the collector connected to the common and the *common base* (CB) configuration with the base connected to the common.

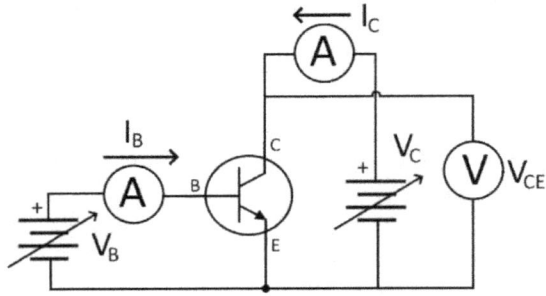

The circuit, as configured above, provides a way to measure the collector current $I_C$ for a range of collector-to-emitter voltage $V_{CE}$ at a given base current $I_B$. Ammeters inserted in the base and collector leads monitor the base current $I_B$ and collector current $I_C$, respectively. By varying voltage source $V_B$, we can set a base current $I_B$ and then measure collector current $I_C$ for a range of collector voltages $V_{CE}$ set by variable voltage supply $V_C$. The figure below shows the family of curves obtained for a popular silicon NPN BJT, the 2N3904.

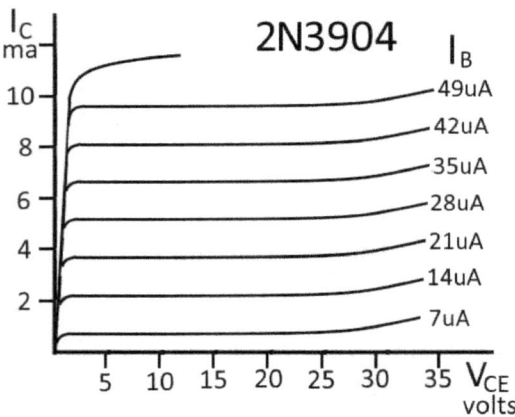

The vertical axis shows a range of operating collector currents from 0 to 10 ma. The horizontal axis shows a range of operating collector voltages from 0 to 35 v. The family of curves represents a range of base currents from 7 to 49 µA. We call this set of curves the *collector characteristics* for the 2N3904.

The figure below shows the value of β (β = $I_C$ / $I_B$) at each value of $I_B$.

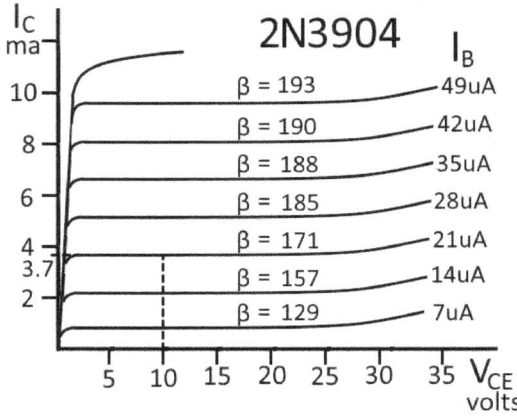

The fact that β decreases as the collector current I_C decreases is common to most BJTs. In large signal applications, β variability introduces amplification nonlinearity and distortion. In addition, if we want to des gn for a specific *DC operating point* (i.e., a specific collector current I_C and V_CE), we must consider which I_B curve to use and the corresponding value of β.

To make matters worse, β is also subject to product variability. For example, the specified β for the 2N3904 ranges from 100 to 300 at 25°C. Such wide variation in β can result in significant collector current and voltage differences. To build an amplifier, we need a circuit design that overcomes β variation by (1) providing the BJT with a predictable and stable DC operating point and (2) lessening the variation in β as a signal varies I_B. To explore this challenge, we consider the basic CE amplifier design below.

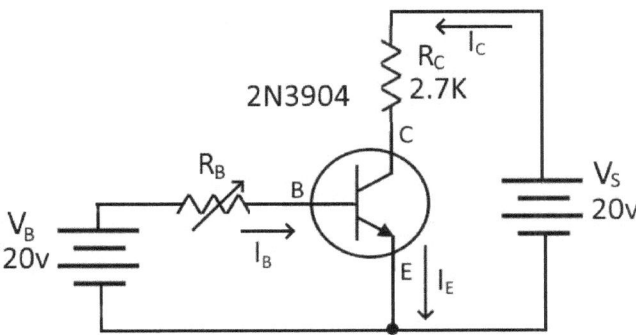

The base bias supply consisting of voltage source V_B and resistor R_B sets the collector current IC to a specific value. As a practical matter, we have chosen V_B equal to $V_S$. If we were to build it, we would want only one external voltage supply. Because the base current $I_B$ is exceedingly small, resistance $R_B$ will be large, more than 100,000 $\Omega$.

First, we adjust $R_B$ so that the value of $I_C$ drops one-half of $V_S$ across collector resistor $R_C$. That way, $V_{CE}$ is free to change ±10 volts as $I_C$ changes from 0 to twice $I_C$'s initially set value. Using Ohm's Law, the required DC operating collector current $I_c$ = 10 / 2700 = 3.7 ma.

From the 2N3904 collector characteristics above β at $I_C$ = 3.7 ma is about 175, corresponding to a base current $I_B$ of 21 µA. The DC operating point would be (10 volts, 3.7 ma). To overcome a BJT's energy gap and begin behaving like a transistor, we must reach a $V_{BE}$ potential of ~0.7 volts for a silicon BJT or ~0.3 volts for a germanium BJT. Using 0.7 volts in our example and noting that $I_B$ = 21µA, we would make $R_B$ = (20 − 0.7) / 21·10$^{-6}$ = 919kΩ.

We will use the *load line* analysis method to explore how the variation of β affects the range of DC operating points. According to circuit analysis (Kirchhoff's Law), the sum of voltages in the collector circuit above must be zero. Thus, we have,

$$20 - V_{R_C} - V_{CE} = 20 - R_C \cdot I_C - V_{CE} = 0$$

$$20 - 2700 \cdot I_C - V_{CE} = 0$$

Rearranging the equation, we have

$$I_C = \frac{1}{2700} \cdot V_{CE} + \frac{20}{2700} \quad \text{(Eqn. 5-5)}$$

Eqn. 5-5 has two unknowns, $I_C$ and $V_{CE}$, so we need two independent mathematical relationships to solve it. Equation 5-5 is one, and the 2N3904 collector characteristics graphically show $I_C$ versus $V_{CE}$ is the other.

Note: Recall from algebra that Eqn. 5-5 is in the standard form Y = mX + B where X is $V_{CE}$, Y is $I_C$, m is the slope (1/2700), and B is the Y-axis (the $I_{CE}$ axis) intercept (20/2700).

We calculate the endpoints of Eqn. 5-5 to plot the load line on the collector characteristics. The upper left endpoint is the value of $I_c$ when $V_{CE}$ is zero. The lower right endpoint is the value of $V_{CE}$ when $I_c$ is zero. We find the first by making $V_{CE}$ zero in Eqn. 5-5 and solving for $I_c$, which is 20 / 2700 = 7.4 ma. When we make $I_c$ zero in Eqn. 5-5, we find $V_{CE}$ is 20 volts.

Drawing a line between the endpoints we have,

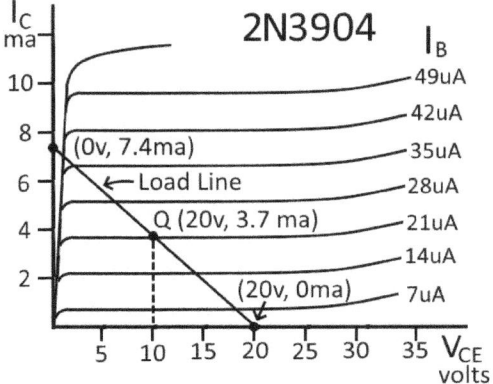

Recall from algebra that the solution to the two equations is their intersection point. Choosing the $I_b$ = 21 µA $I_c$ curve gives us the desired DC operating point, $V_{CE}$ = 10 volts, and $I_c$ = 3.7 ma.

The point labeled "Q" above is the desired DC operating point or *quiescent point*. Hereafter, we will use "quiescent point" when refe ring to the DC operating point when no signal is present.

If we vary $I_B$ above and below 21 µA, the quiescent point will move up and down the load line. We see that the change in $I_c$ is a few milliamps while the $I_B$ change is only a few microamps. Thus, the circuit achieves current amplification, as we proposed at the outset. Though not apparent, $V_{CE}$ is changing in volts while $V_{BE}$ is changing in millivolts. This indicates we have voltage gain as well. In fact, since power gain is the product of voltage and current gain, the BJT amplifier is also exhibiting power gain. That is, microwatts of input power produce milliwatts of output power.

Because we assumed β was ~170, we achieved the quiescent point we wanted. But we know from the product variability issue raised earlier that we cannot depend on β being 170 "off the shelf."

We will now use load line analysis to explore "graphically" what happens when β varies from 100 to 300.

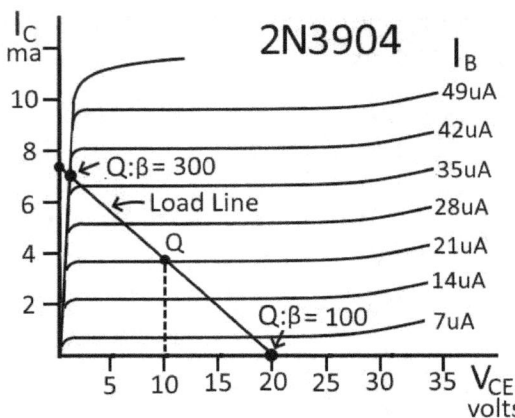

If we hold $I_B$ to 21 μA and vary β, the quiescent point moves to opposite ends of the load line. β = 300 drives the BJT into *saturation,* where the collector voltage drops to near zero. For β = 100, the BJT is in *cutoff* with negligible collector current, and $V_{CE}$ is equal to the collector supply voltage. Neither situation permits the BJT to function as an amplifier.

Note: Variation in product β changes the collector characteristics itself. For β = 300, the $I_B$ curves spread out so that $I_B$ = 21 μA falls just above the $I_B$ = 35 μA curve in the original collector characteristics. A β = 100 has the opposite effect; the $I_B$ curves compress, and $I_B$ = 21 μA falls near the $V_{CE}$ axis.

Restricting BJTs to a narrow range of βs in mass production is not practical. So, we must consider other ways to stabilize the quiescent point.

In the scenario above, we used a large value of $R_B$ (919 KΩ). The result is that the small (few millivolts) changes in base-emitter voltage from β variation had no significant effect on $I_B$. In effect, this combination of large $V_B$ and large $R_B$ acted like a current source, pumping a fixed current (21 μA) into the BJT's base regardless of the value of $V_{BE}$.

If we had used much smaller values of $V_B$ and $R_B$, the change in $V_{BE}$ would have varied the value of $I_B$ to compensate for β variation. Lowering $V_B$, however, would defeat our intent to use a single power supply. It would appear then that to make a practical BJT amplifier that overcomes β variability; we must develop a different base biasing technique.

In the next chapter, we consider a design approach to do that precisely!

# Chapter 6 – BJT Basic Bias Design

In the previous chapter, we saw BJT's potential for use as an amplifier. We also found that we must utilize a design that provides a *stable quiescent point*. If the BJT's β varies too widely (as it does "off the shelf"), the quiescent point can move to the ends of the load line, making the amplifier dysfunctional. In this chapter, we describe a bias technique that satisfactorily addresses the product variability of β and keeps $I_C$ within specified limits.

Note: Operating temperature also affects the collector current $I_C$ and the quiescent point. Temperature instability can result in BJT self-destruction! The technique we use to stabilize β will also address thermal instability.

In the figure below, we have revised the simple biasing circuit of the previous chapter by showing a single voltage supply.

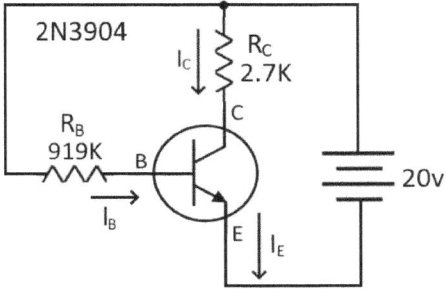

As shown, the 20-volt source supplies both the base and collector of the 2N2904. Recall that we calculated the value of resistor $R_B$ to provide the 21 µA base current $I_B$ that, based on β = 171, would produce the required collector current $I_C$ of 3.7 ma and a $V_{CE}$ of 10 volts.

We first study the control aspect of this simple bias approach by considering what happens to $V_{BE}$ when β increases. As β increases, so does $I_c$. In the case of silicon BJTs like the 2N3904, a 9 ma rise in $I_c$ produces a 100 mv rise in $V_{BE}$, or $11.1\frac{mv}{ma}$. Suppose β above changes to the worst case 300, $I_c$ will increase from 2.6 ma to 6.3 ma. To compensate, $I_B$ would have to decrease by 8.7 µA to maintain $I_C$ at 3.7 ma. Given the 919 KΩ resistance, the decrease in $I_B$ would be $\frac{11.1 \cdot 2.6 \cdot 10^{-3}}{919 \cdot 10^3} = {\sim}0.031$ µA. While this corrective action is in the right direction to reduce $I_c$, it is not close to the 8.7 µA needed to bring $I_c$ back down to the 3.7 ma.

Note: We refer to this corrective action as *negative feedback*. The increase in β produces a corrective action that attempts to bring the circuit back to the desired quiescent point. P*ositive feedback, on* the other hand, would apply an action that caused the circuit to move farther away from the desired quiescent point. In general, negative feedback has a stabilizing effect, while positive feedback is destabilizing. In the situation above, the large value of $R_B$ fails to supply sufficient negative feedback to correct the change in Ic fully.

To better regulate $V_{BE}$, we need to lower the base supply voltage (call it $V_B$) and the value of $R_B$. A voltage divider network like that shown in Figure A below is a step in the right direction.

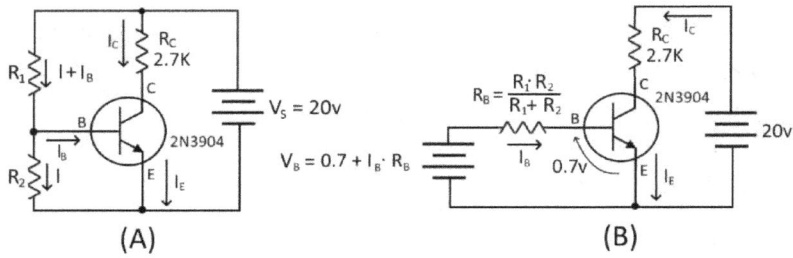

$$R_B = \frac{R_1 \cdot R_2}{R_1 + R_2} \quad \text{(Equ. 6-1a)}$$

If we choose $R_1 + R_2$ so their current I is ten or more times larger than base current $I_B$, changes in $I_B$ will insignificantly affect the voltage divider output. Then, ignoring $I_B$, we calculate values for $R_1$ and $R_2$ that give the desired $V_{BE}$.

Circuit B above shows the Thevenin Equivalent of the voltage divider. (For more information on the Thevenin Equivalent and the Thevenin Theorem, see Appendix H.) $R_B$ is the parallel combination of $R_1$ and $R_2$. $V_B$ is larger than $V_{BE}$ by the voltage drop across $R_B$. Therefore, the controlling equations are,

$$R_B = \frac{R_1 \cdot R_2}{R_1 + R_2} \quad \text{(Equ. 6-1a)}$$

$$V_B = 0.7 + R_B \cdot I_B \quad \text{(Equ. 6-1b)}$$

Using the 2N3904 parameters β = 175, $V_{BE}$ = 0.7 volts, $I_C$ = 3.7 ma, and $I_B$ = 21 · $10^{-6}$ ma, we first set I = 10 · 21 · $10^{-6}$ = 210 · $10^{-6}$. Assuming $V_{BE}$ = 0.7 volts,

$$R_2 = \frac{0.7}{210 \cdot 10^{-6}} = 3333 \ \Omega$$

$$R_1 = \frac{(20 - 0.7)}{210 \cdot 10^{-6}} = 91{,}905 \ \Omega$$

$$R_B = \frac{3333 \cdot 91{,}905}{3333 + 91{,}905} = 3216 \ \Omega$$

$$V_B = 0.7 + 3216 \cdot 21 \cdot 10^{-6} = 0.767 \ v$$

To check, we calculate $I_B$ = (0.767 – 0.7) / 3216 = 0.067 / 3205 ≈ 21 µA.

Here, then, is the equivalent circuit.

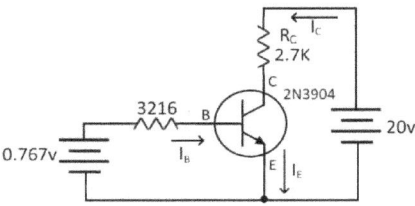

The new $R_B$ 3216 $\Omega$ is much smaller than the original $R_B$ 919 K $\Omega$, and the source voltage $V_B$ is closer to 0.7 volts. This combination should act less like a constant current source and might provide the negative feedback we need to stabilize $I_C$. Recalculating the correcting current, we have $\frac{11.1 \cdot 2.6 \cdot 10^{-3}}{3216} \approx 9.0$ µA, close to the 8.7 µA needed. While this looks good, what happens at the other extreme of β for the 2N3904, β = 100? In this case, $I_C$ drops to $100 \cdot 21 \cdot 10^{-6} = 2.1$ ma. To bring $I_C$ back to 3.7 ma, $I_B$ must increase by $\frac{1.6 \ ma}{100} = 16$ µA. Calculating the correcting current, we have $\frac{11.1 \cdot 1.6 \cdot 10^{-3}}{3216} = 5.5$ µA, short of the 16 µA needed.

We could lower $R_B$ still further, but at some point, the load on the preceding stage might become too great. While a voltage divider is a step in the right direction, it cannot provide sufficient negative feedback. We must look for another source of negative feedback.

Consider the BJT model circuit below.

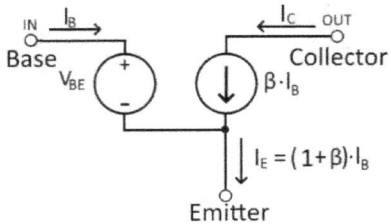

The circuit introduces two new symbols. On the base side, the circle symbol with the enclosed plus and minus signs is an *ideal voltage source* that provides the base-emitter potential $V_{be}$, approximately 0.7 volts for silicon BJTs and 0.3 volts for germanium BJTs. An *ideal voltage source* can supply the indicated voltage regardless of the current we draw from it. We use this symbol primarily to represent internally generated voltages, such as in the model above. We use the familiar battery symbol.

It represents an ideal voltage source <u>external</u> to the circuit. It could represent an actual battery or a DC power supply.

The circle symbol on the collector's side with an arrow inside represents an *ideal current source*. It supplies the indicated current in the direction of the arrow regardless of what voltage appears across its terminals. Current sources may seem strange, but we have encountered one in Chapter 5, the BJT. Consider again the 2N3904 collector characteristics below.

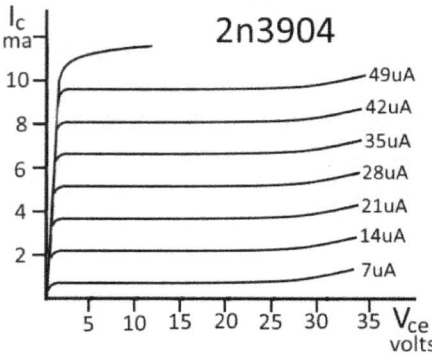

Consider the collector current $I_C$ for a given base current $I_B = 28\ \mu A$. Over the range of voltage $V_{CE}$ (5 to 25 volts), the collector current does not vary significantly from 5.1 ma. Therefore, the collector and emitter terminals of the 2N3904 function as a current source of value $I_C = \beta \cdot I_B$ for a given range of $V_{CE}$.

One other attribute of ideal voltage and current sources is their value can be dependent on another circuit parameter. In the model above, the ideal current source's value depends on the base current $I_B$, i.e., $I_C = \beta \cdot I_B$. Note that this modeling is consistent with the collector characteristics above.

Looking at the circuit, the emitter current is composed of $\beta \cdot I_B$ and $I_B$ so we can write it as

$I_E = I_B + I_C = I_B + \beta \cdot I_B = (1 + \beta) \cdot I_B$ (Eqn. 6-2)

This simple BJC model only considers that collector current $I_c$ is proportional to the base current $I_B$ ($I_C = \beta \cdot I_B$). It does not, for example, consider other factors, such as $V_{BE}$'s dependency on $I_C$. We will address this and other limitations presently, but for now, we will ignore them.

Consider our base bias circuit, to which we have added a resistor $R_E$ from emitter to common.

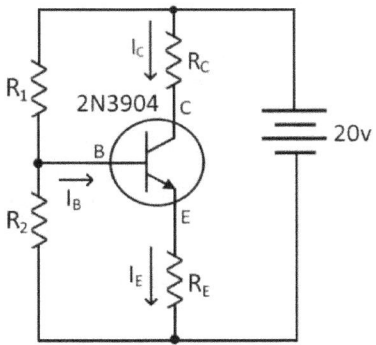

The voltage drop across $R_E$ varies directly with $I_C$ and, more importantly, with $\beta$. As $\beta$ increases (or decreases), $I_C$ increases (or decreases), and the voltage drop across $R_E$ increases (or decreases).

To study if the $R_E$ voltage drop will provide the additional negative feedback we need, consider the circuit above with 2N3904 BJT replaced by our BJT model.

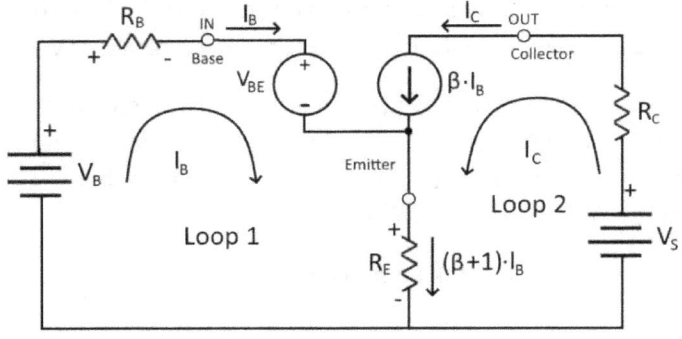

Also, we replaced the voltage divider bias circuit with its Thevenin Equivalent. Writing the voltage loop equation for the base circuit, Loop 1, we have

$$V_B - R_B \cdot I_B - V_{BE} - (\beta + 1) \cdot I_B \cdot R_E = 0$$

Solving for $I_B$, we have

$$I_B = \frac{V_B - V_{BE}}{R_B + (\beta + 1) \cdot R_E} \quad \text{(Eqn. 6-3)}$$

If we assume that $(\beta + 1) \cdot R_E \gg R_B$, then.

$$I_B = \frac{V_B - V_{BE}}{(\beta + 1) \cdot R_E}$$

If β increases, the denominator becomes larger, and $I_B$ decreases. The opposite also follows; a decreased β produces an increased $I_B$. In either case, a change in $I_C$ produces a corrective change in $I_B$. Thus, we have the additional negative feedback we desired. By choosing a value for $R_E$ appropriately, we can provide additional compensation for changes in β and reduce variations in $I_C$.

$R_E$ also provides corrective action for other factors that change collector current, such as increased operating temperature. We will also find that it reduces distortion due to changes in β over large signal excursions of $I_B$. Introducing $R_E$ also takes the pressure off choosing a small value of $R_B$. We can achieve the needed level of correction without loading down a preceding stage,

To continue our analysis, we consider the collector circuit Loop 2. The loop equation is,

$$V_S - I_C \cdot R_C - V_{CE} - (I_C + I_B) \cdot R_E = 0 \quad \text{(Eqn. 6-4)}$$

Since $I_C \gg I_B$, $I_C + I_B \approx I_C$ and,

$$I_C = \frac{1}{R_L + R_E} \cdot V_{CE} + \frac{V_S}{R_L + R_E} \quad \text{(Eqn. 6-5)}$$

Eqn. 6-5 should be familiar. Its form is the same as Eqn. 5-5 that we used to find the quiescent point for the example circuit. If $R_E = 0$, the equations are the same. The similarity might tempt us to use the load line method of Chapter 5 as the basis of a design approach. This is not a promising idea because the wide-ranging variability of β variability means there is not a single collector characteristic valid for all BJTs of a particular type.

We need an equation that considers variations in β and the allowable range of acceptable values of $I_C$. An excellent place to start is Eqn. 6-3 that gives us $I_B$. Knowing that $I_C = \beta \cdot I_B$, we can write,

$$I_C = \frac{\beta \cdot (V_B - V_{BE})}{R_B + (\beta + 1) \cdot R_E} \quad \text{(Eqn. 6-6)}$$

Define $\beta_L$ as the smallest expected value of β, and $\beta_H$ as the highest. Similarly, assume $I_{CL}$ is the lowest expected value of collector current when $\beta = \beta_L$. Likewise, define $I_{CH}$ when $\beta = \beta_H$. Eqns. 6-7 and 6-8 establish the relationship between these values.

$$I_{CL} = \frac{\beta_L \cdot (V_B - V_{BE})}{R_B + (\beta_L + 1) \cdot R_E} \quad \text{(Eqn. 6-7)}$$

$$I_{CH} = \frac{\beta_H \cdot (V_B - V_{BE})}{R_B + (\beta_H + 1) \cdot R_E} \quad \text{(Eqn. 6-8)}$$

Next, we take the ratio of these equations.

$$\frac{I_{CH}}{I_{CL}} = \frac{\beta_H}{\beta_L} \cdot \frac{R_B + (\beta_L + 1) \cdot R_E}{R_B + (\beta_H + 1) \cdot R_E} \quad \text{(Eqn. 6-9)}$$

For our design procedure, we will specify the $I_C$s and βs. If we solve Eqn. 6-9 for $R_B/R_E$, then that will give us a value on which to begin the design. Solving Eqn. 6-9 for $\frac{R_B}{R_E}$, we have

$$\frac{R_B}{R_E} = \frac{(\beta_L + 1) - k \cdot (\beta_H + 1)}{(k - 1)} \quad \text{where } k = \frac{I_H \cdot \beta_L}{I_L \cdot \beta_H} \quad \text{(Eqn. 6-10)}$$

We will also need equations for $V_B$ and $R_B$.

$$V_B = \frac{R_2}{R_1 + R_2} \cdot V_S \quad \text{(Equ. 6-11a)}$$

$$R_B = \frac{R_1 \cdot R_2}{R_1 + R_2} \quad \text{(Equ. 6-11b)}$$

If $V_B$ and $R_B$ are known, Eqns. 6-11a and b solved simultaneously yield values for $R_1$ and $R_2$.

$$R_1 = \frac{V_S \cdot R_B}{V_B} \quad \text{(Equ. 6-12a)}$$

$$R_2 = \frac{R_1 \cdot R_B}{R_1 - R_B} \quad \text{(Eqn. 6-12b)}$$

That should do it.

To verify that our method reduces β variation effects on collector current, we will use this circuit.

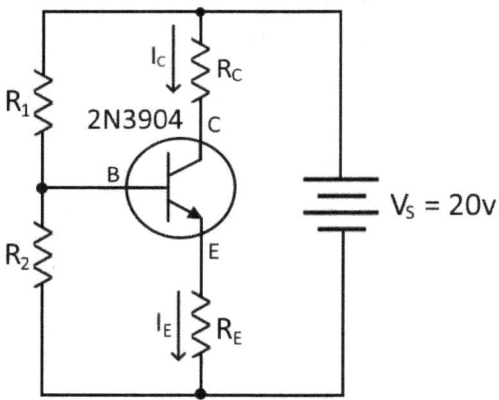

Here are the design parameters:

$V_S$ = 20 volts
$V_{CE}$ = 10 volts
$V_{BE}$ = 0.7 volts
$R_B$ = 10 KΩ
$I_{CL}$ = 3.5 ma; $I_{CH}$ = 3.9 ma

$I_C$ (nominal) = 3.7 ma
$\beta_L$ = 100; $\beta_H$ = 300

As noted before, we are free to choose a larger value of $R_B$ or choose a value of $R_B$ to suit a particular application.

Step 1 – We use Eqn. 6-10 to calculate $R_B/R_E$.

$$k = \frac{I_H \cdot \beta_L}{I_L \cdot \beta_H} = \frac{.0039 \cdot 100}{.0035 \cdot 300} = 0.3714$$

$$\frac{R_B}{R_E} = \frac{(100 + 1) - 0.3714 \cdot (300 + 1)}{(0.3714 - 1)} = \frac{-10.791}{-0.6286} = 17.2$$

Substituting $R_B$ = 10 KΩ, we find that $R_E$ = 10000/17.2 = 581 Ω and select a standard value 560 Ω resistor.

Step 2 – We calculate $R_L$. We assume that $I_E = I_C$, then from Eqn. 6-4, we calculate $V_{CE}$.

$$V_{CE} = V_s - (R_C + R_E) \cdot I_C \text{ and } R_C = \frac{V_S - R_E \cdot I_C - V_{CE}}{I_C}$$

Substituting values, we find that,

$$R_C = \frac{20 - 560 \cdot 0.0037 - 10}{0.0037} = 2142 \ \Omega$$

and use a standard value of 2200 Ω.

Step 3 – We calculate $V_B$ using Eqn. 6-6.

$$I_C = \frac{\beta \cdot (V_B - V_{BE})}{R_B + (\beta + 1) \cdot R_E} \text{ and } V_B = \frac{(R_B + (\beta + 1) \cdot R_E) \cdot I_C}{\beta} + V_{BE}$$

Substituting values, we find that,

$$V_B = \frac{(10000 + (300 + 1) \cdot 560) \cdot 0.0039}{300} + 0.7 = 3.02 \text{ volts}$$

Step 4 – We calculate $R_1$ and R2 using Eqn. 6-12a and Eqn. 6-12b.

$$R_1 = \frac{20 \cdot 10000}{3.02} = 66,225 \ \Omega$$

$$R_2 = \frac{66225 \cdot 10000}{66225 - 10000} = 11,778 \, \Omega$$

Using the nearest standard values, we choose $R_1$ = 68 KΩ and $R_2$ = 12 KΩ.

The final circuit is,

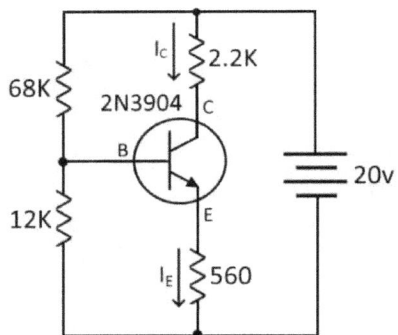

Note: From now on, we will employ two methods to verify circuit operation. In most cases, we will use circuit simulation, as described in the introduction. Sometimes, we will breadboard the circuit to verify its performance under actual operating conditions.

The circuit test gave the following results:

|  | | Simulation | | Breadboard | |
|---|---|---|---|---|---|
|  | Expected | With β = 100 | With β = 300 | With β = 110 | With β = 270 |
| $I_C$ | 3.7±0.2 ma | 3.3 ma | 3.7 ma | 3.3 ma | 3.4 ma |

While the values were off slightly, more importantly, the collector currents varied no more than 0.4 ma over a wide range of β. By adding $R_E$, we verified our design method achieved the desired quiescent point stability.

Note: To further verify the improvement made with $R_E$, we removed it from the breadboard and adjusted $R_1$ and $R_2$ so that $I_C$ = 3.7 ma. We then tried two BJTs with βs of 110 to 270, respectively. Without $R_E$, the collector current $I_C$ varied from 2.1 ma to 5.0 ma!

Thus far, we have not considered the effects of temperature on collector current. Leakage current, β, and $V_{BE}$ are temperature dependent and can affect collector current $I_C$.

In the next chapter, we consider these temperature effects.

# Chapter 7 – Basic Temperature Effects

In this chapter, we will develop stability factors that we can use to evaluate temperature effects on our amplifier design. These stability factors will quantify variation in collector current $I_C$ due to the temperature dependency of leakage current $I_{CEO}$, base-emitter voltage $V_{BE}$, and $\beta$.

Regarding temperature dependency, here are three rules of thumb:

1. Leakage current doubles with about every 10°C rise in temperature.

2. $V_{BE}$ decreases by about 2.5 mv for every 1°C rise in temperature.

3. $\beta$ increases 5-7% for every 10°C increase in temperature.

Environmental heating and self-heating affect the collector current. Defining *stability factors* gives us a way to estimate these thermal effects and avoid heating-induced problems.

## Leakage Stability Factor – S

The definition of the *leakage stability factor* S is the partial derivative of $I_C$ with respect to leakage current $I_{CEO}$, which we can write as,

$$S = \frac{\partial I_C}{\partial I_{CEO}}$$

$I_{CEO}$ is the collector-emitter leakage current with the base open-circuited. We can find S by taking the regular derivative of $I_C$ with respect to $I_{CEO}$ while holding $\beta$ and $V_{BE}$ constant. Said a less mathematical way, S is the change in collector current $I_C$ ($\Delta I_C$) for a slight change in leakage current ($\Delta I_{CEO}$) while holding $\beta$ and $V_{BE}$ constant; that is,

$$S \approx \frac{\Delta I_C}{\Delta I_{CEO}} \text{where, } \Delta I_{CEO} \text{ is a slight change in } I_{CEO}. \text{ (Eqn.7-1)}$$

Another commonly specified leakage current is $I_{CO}$. It is the leakage base current when the base is open-circuited. As with $I_{CEO}$, it produces collector current. We can show that,

$$I_{CEO} \approx (\beta + 1) \cdot I_{CO} \ (Eqn. 7 - 2a) \text{ and } \Delta I_{CEO} \approx (\beta + 1) \cdot \Delta I_{CO} \ (Eqn. 7 - 2b)$$

We will use these relationships shortly.

The usual approach is to determine S for a particular circuit configuration (CE, CC, or CB). Shown below is the BJT model we will use.

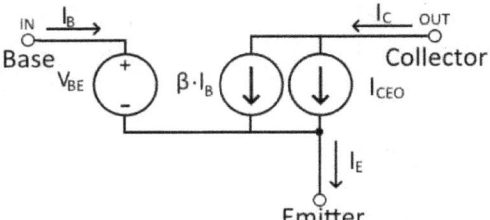

It is like the hybrid model with the addition of a current source representing leakage current $I_{CEO}$. The figure below shows a CE amplifier circuit with this model in place of the BJT.

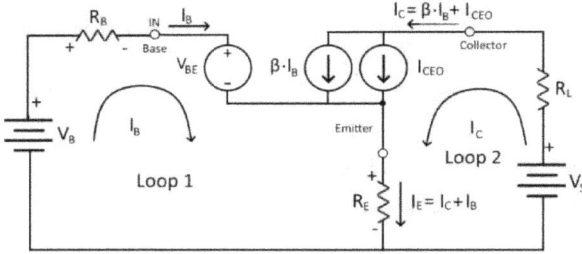

The total collector current is,

$$I_C = \beta \cdot I_B + I_{CEO} \quad \text{(Eqn. 7-3)}$$

The following three equations follow from the circuit:

$$V_B - V_{BE} = R_B \cdot I_B + R_E \cdot (I_B + I_C) \quad \text{(Eqn. 7-4)}$$

From Eqn. 7-3,

$$I_B = \frac{I_C - I_{CEO}}{\beta} \quad \text{(Eqn. 7-5)}$$

When we substitute Eqn. 7-5 into Eqn. 7-4, we obtain,

$$V_B - V_{BE} = R_B \cdot \frac{I_C - I_{CEO}}{\beta} + R_E \cdot \frac{I_C - I_{CEO}}{\beta} + R_E \cdot I_C \quad \text{(Eqn. 7-6)}$$

Solving for $I_C$, we find that,

$$I_c = \frac{\beta \cdot (V_B - V_{BE})}{R_B + (\beta + 1) \cdot R_E} + \frac{R_B + R_E}{R_B + (\beta + 1) \cdot R_E} \cdot I_{CEO} \quad \text{(Eqn. 7-7)}$$

Assume a slight change in $\Delta I_{CEO}$ produces a change in $I_C$. We can represent this change in Eqn. 7-7 as,

$$I_c + \Delta I_c = \frac{\beta \cdot (V_B - V_{BE})}{R_B + (\beta + 1) \cdot R_E} + \frac{R_B + R_E}{R_B + (\beta + 1) \cdot R_E} \cdot (I_{CEO} + \Delta I_{CEO}) \quad \text{(Eqn. 7-8)}$$

Subtracting Eqn. 7-7 from Eqn. 7-8 and dividing by $\Delta I_{CEO}$, we find that,

$$I_C = \frac{R_B + R_E}{R_B + (\beta + 1) \cdot R_E} \cdot \Delta I_{CEO}$$

Then, using Eqn. 7-2b, we have,

$$S \approx \frac{\Delta I}{\Delta I_{CO}} = \frac{(\beta + 1) \cdot (R_B + R_E)}{R_B + (\beta + 1) \cdot R_E} \quad (Eqn.\,7 - 9)$$

Rearranging, we obtain,

$$S = \frac{(R_B + R_E)}{\dfrac{R_B}{(\beta + 1)} + R_E} \quad \text{(Eqn. 7-10)}$$

After we rearrange Eqn. 7-9, we have,

$$\Delta I_C \approx S \cdot \Delta I_{CO} \quad (Eqn.\,7 - 11)$$

From this relationship, we conclude that the smaller S is, the less effect leakage current will have on collector current. For BJTs, an S less than 25 proves satisfactory.

We can check the stability factor of the CE amplifier in Chapter 6. See the CE circuit below.

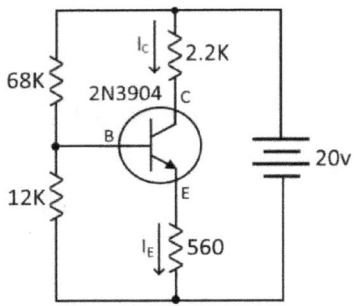

For this example, we assumed $R_B = 10\ K\Omega$ and calculated $R_E = 560\ \Omega$.

$$S = \frac{(10000 + 560)}{\dfrac{10000}{(300 + 1)} + 560} = 17.8$$

S is less than 25, so we can assume that our design is stable with respect to leakage current.

In addition, we can calculate the collector current change over the operating temperature range, which we will assume is up to $70°\,C$. From 2N3904 specs, we find $I_{CO} = 50$ nA at $25°\,C$. To find $I_{CO}$ at $70°\,C$, we use the chart below.

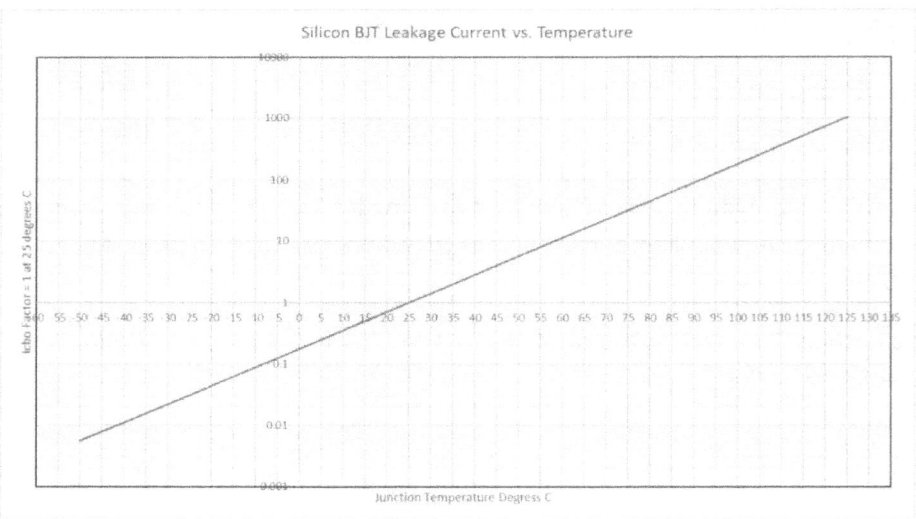

The factor at $70°\,C$ is approximately 22, so $I_{CO}$ is $22 \cdot 50$ nA = 1.1 µA, and $\Delta I_{CO}$ is 1.1 µA − 0.05 µA = 1.05 µA. We then calculate the change in current using Eqn.

7-11; $\Delta I_C$ = 17.8 · 1.05 μA = 0.019 ma, which will have a negligible effect on the quiescent collector current 3.7 ma.

Had we used a germanium BJT with $I_{CO}$ = 6 μA, $I_C$ would have increased by 6 · 21 · 17.8 = 2.24 ma. This would drive the BJT into saturation, making the CE amplifier inoperative. This supports our contention that silicon BJTs are far superior to germanium BJTs.

## $V_{BE}$ Stability Factor- $S_V$

Next, we will check stabilization with respect to $V_{BE}$. We start with Eqn. 7-7,

$$I_C = \frac{\beta \cdot (V_B - V_{BE})}{R_B + (\beta + 1) \cdot R_E} + \frac{R_B + R_E}{R_B + (\beta + 1) \cdot R_E} \cdot I_{CEO}$$

We first define $S_V$, the base-emitter voltage $V_{BE}$ stability factor, below.

$$S_V = \frac{\delta I_C}{\delta V_{BE}} \approx \frac{\Delta I_C}{\Delta V_{BE}} \quad \text{where, } \Delta V_{BE} \text{ is small. (Eqn. 7-12)}$$

Following a similar procedure to that we used to find stability S, we find that,

$$S_V = \frac{-\beta}{R_B + (\beta + 1) \cdot R_E} \quad \text{(Eqn. 7-13)}$$

and

$$\Delta I_C \approx S_V \cdot \Delta V_{BE} \quad \text{(Eqn. 7-14)}$$

Using the values $\beta$ = 175, $R_E$ =β 10000, and $R_B$ = 560, $S_V$ is,

$$S_V = \frac{-175}{10000 + (172) \cdot 560} = -0.00161 \frac{ma}{mv}$$

Once again, we use 70°C as the upper temperature. $V_{BE}$ decreases 2 mv per °C, so we find that,

$$\Delta V_{BE} = (70 - 25) \cdot (-2) = -90 \, mv$$

Using Eqn. 7-14,

$$\Delta I_C \approx (-00161) \cdot (-90) = 0.147 \, ma$$

Adding the $I_C$ increase due to leakage, the total change is ~0.17 ma, within the 0.2 ma limit we imposed on the design. We conclude that our design is stable for temperature effects on $V_{BE}$.

## β Stability Factor – $S_\beta$

In our Chapter 6 design, we accounted for a range of β values (100 to 300, to be exact), so the final calculation of β stability $S_\beta$ is superfluous. Nevertheless, having a way to calculate $S_\beta$ can come in handy if we want to check $S_\beta$ for an existing design.

The derivation of $S_\beta$ is tedious and beyond the scope of this book. If β is significantly greater than 1 (β >> 1), then,

$$S_\beta \approx \frac{\Delta I_C}{\Delta \alpha} \approx \frac{S \cdot I_Q}{\alpha} \text{ where } \alpha = \beta / (\beta + 1). \quad (\text{Eqn. 7-15})$$

Using the example values above,

$$S_\beta \approx \frac{17.8 \cdot .0037}{0.9942} = 0.0662 \text{ A}$$

Assuming β changes from 100 to 300, the change in collector current is

$$\Delta I_C = S_\beta \cdot \Delta \alpha = 0.0662 \cdot 0.00658 = 0.44 \, ma$$

Recall that we designed our amplifier for a ±0.2 ma variation (0.40 ma total) in $I_C$ with a β range of 100 to 300. Given all the approximations inherent in our calculations, the small difference 0.40 ma vs. 0.44) is acceptable. The point here is that for our design, product variation in β will not have a significant effect on the quiescent collector current $I_C$.

We also must check on the increase in β with temperature. For a temperature rise from 25°C to 70°C, we can estimate that the change in β will Δβ = ((70 – 25) / 10) · 5 = 22.5. For β = 100 (worse case), Δα = 0.0018 and the change in collector current is,

$$\Delta I_C = S_\beta \cdot \Delta \alpha = 0.0662 \cdot 0.0018 = 0.12 \, ma$$

As the initial β increases, the change in collector current decreases. In this example, at β = 300, $\Delta I_C$ = 13 μA is a negligible value. If we consider this along with the broader allowable value of $\Delta I_C$ at lower values of $I_C$, β temperature variation presents no problem with our design.

Based on the author's experience, the design method of Chapter 6 yields CE amplifiers that are stable for the temperature range of 0°C to 70°C, which is a consumer product standard.

With an excellent static design for our amplifier (i.e., a stable quiescent point), we are ready to explore small signal amplification in the next chapter.

# Chapter 8 – Small Signal BJT Amplifiers

We choose the AC analysis method for BJT amplifiers based on the size of the AC signal. Our two choices are *small signal analysis* and *large signal analysis.* Small signal analysis assumes the AC signal is sinusoidal and does not deviate more than 10% from the BJT's quiescent collector current. In this case, we use a *small signal model* to represent the BJT and then apply standard circuit analysis methods to determine performance characteristics such as voltage and current gain. For large signal analysis, where signals deviate beyond 10% of the BJT's quiescent collector current, we use load line analysis to obtain voltage and current gain. In this chapter, we address small signal analysis. In a later chapter, we will address large signal analysis.

## Ideal Amplifiers

We represent an *ideal amplifier* by the block figure below.

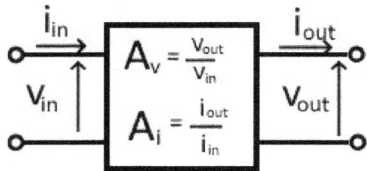

We use lowercase letters for voltage and current to represent small signal values. An *ideal amplifier* has the following performance characteristics:

Voltage Gain $A_v = \dfrac{v_{out}}{v_{in}}$   (Eqn. 8-1a)

Current Gain $A_i = \dfrac{i_{out}}{i_{in}}$   (Eqn. 8-1b)

The amplifier is" ideal" in that gains are not dependent on the frequency, size, or type of signal. In addition, these gains are unaffected by what we connect to the amplifier's input or output. For example, suppose we connect two ideal amplifiers in tandem (we call this *cascading*), as shown below.

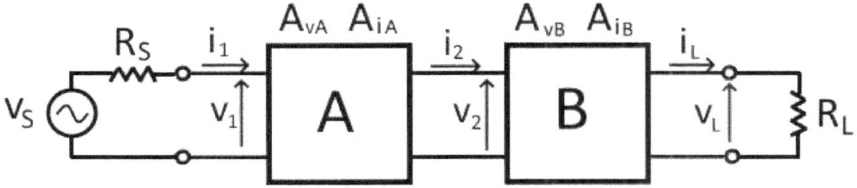

Because A and B are ideal amplifiers, input and output resistances (like $R_S$ and $R_L$) do not affect their gains. In addition, the cascaded gains are the simple products of individual gains, as shown below.

$$A_v = \frac{v_L}{v_1} = A_{vA} \cdot A_{vB} = \frac{v_2}{v_1} \cdot \frac{v_L}{v_2} \quad \text{(Eqn. 8-1c)}$$

$$A_i = \frac{i_L}{i_1} = A_{iA} \cdot A_{iB} = \frac{i_2}{i_1} \cdot \frac{i_L}{i_2} \quad \text{(Eqn. 8-1d)}$$

### Actual Amplifiers

*Actual amplifiers* differ from ideal amplifiers in these ways.

(1) Voltage and current gain are frequency dependent.

(2) The size of the input signal affects voltage and current gain.

(3) The shape of the output signal is not identical to that of the input signal.

(4) External circuits affect voltage and current gain.

(5) Relationships can exist between input and output voltage and currents. For example, an actual amplifier's output voltage could affect its input current, which, in turn, could affect its current gain.

Accounting for these differences is complicated. We can partially address them by adding input and output impedances to the ideal amplifier. See the figure below.

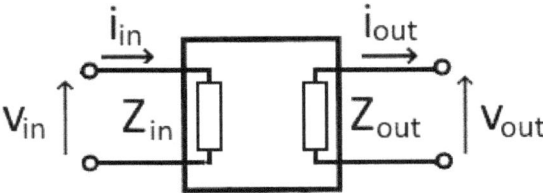

Impedance is a component or circuit's opposition to electron flow. It consists of two components: DC resistance and AC reactance. DC resistance obeys Ohm's Law and is frequency-independent. AC reactance, on the other hand, obeys Ohm's Law but with a frequency dependency.

For a simple example of an input/output interaction, consider what happens when we cascade actual amplifiers, as shown below.

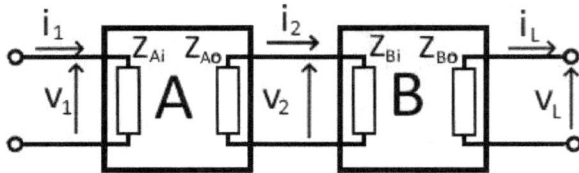

Paralleling A's output impedance with B's input impedance lowers A's effective output impedance. This additional load can, in turn, affect A's voltage and current gain.

To take the next step in accounting for the difference between ideal and actual amplifiers, we make these assumptions.

(1) The amplifier's voltage and current gains are not dependent on the frequency. Later, we will remove this assumption and account for frequency effects.

(2) Input and output impedances are purely resistive, as illustrated below.

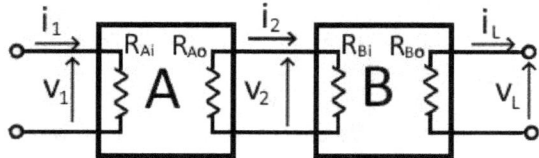

(3) Voltage and currents are linearly related.

(4) The coupling between actual amplifiers is one of three circuit types:

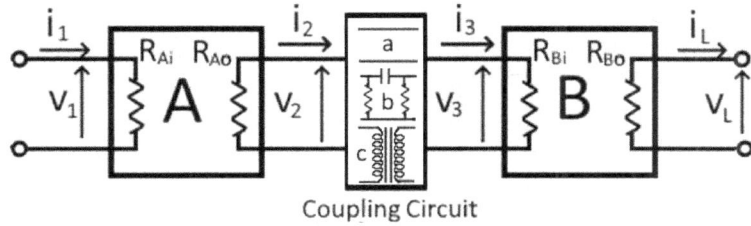

Coupling Circuit

Example "a" is *direct coupling* in which we connect the first amplifier's output directly to the second amplifier's input. This is the optimum choice as it typically minimizes gain and frequency effects.

Example "b" is *resistive-capacitive coupling (RC coupling), which is* necessary to DC isolate actual amplifier outputs and inputs. For example, if the output circuit of amplifier A is at a different DC potential than the input of amplifier B, the capacitor provides the necessary isolation. The resistors shown may be part of the existing circuit to which we add the appropriately sized capacitor. Because the capacitor's signal resistance (called *capacitive reactance*) increases with decreasing frequency, RC couplings limit an actual amplifier's low-frequency response. We will address this in a later chapter.

Example "c" is *transformer coupling (XFRM coupling),* which also DC isolates the two amplifiers. Besides providing isolation, a transformer can also match the input load impedance of the amplifier B with the output impedance of the amplifier A. By doing so, we decrease the loading effect of A on B and maximize power transfer when that is the design focus. A drawback of XFRM coupling is that transformers can introduce frequency effects on gain at low and high signal frequencies.

Note: While we will not discuss it in this book, we use XFRM coupling in radio frequency (RF) amplifiers by tuning the windings of the transformer to resonate at a desired RF frequency. The resulting circuit passes the desired frequency and attenuates frequencies above or below. The result is a *tuned amplifier*.

In the following chapters, we will provide audio amplifier examples of all three couplings, along with their effects on gain and frequency response.

## The Hybrid Model

To analyze our actual amplifier, we selected a small signal model to represent the BJT. While a variety of BJT models exist, we chose the *hybrid model* shown below for our small signal analysis.

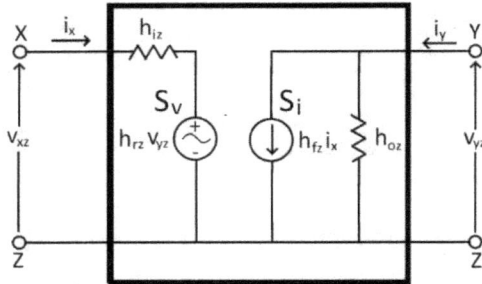

All values are lowercase to reflect they are small signal values (within ±10% of the quiescent collector current). In addition to familiar input and output resistances ($h_{iz}$ and $h_{oz}$), the model includes two dependent sources: voltage source $S_v$ and current source $S_i$. They depend on the values of output voltage $v_{yz}$ and input current $i_x$, respectively. Hybrid parameters $h_{rz}$ and $h_{fz}$ are the respective proportionality constants.

These two linear equations mathematically represent the hybrid model.

$$v_{xy} = h_{iz} \cdot i_x + h_{rz} \cdot v_{yz} \quad \text{(Eqn. 8-2a)}$$

$$i_y = h_{fz} \cdot i_x + h_{oz} \cdot v_{yz} \quad \text{(Eqn. 8-2b)}$$

where the h-parameters are,

$$h_{iz} \approx \frac{\Delta v_{xz}}{\Delta i_x} \text{ where } v_{yz} = 0 \quad \text{(Eqn. 8-3a)}$$

$$h_{rz} \approx \frac{\Delta v_{xz}}{\Delta v_{yz}} \text{ where } i_x = 0 \quad \text{(Eqn. 8-3b)}$$

$$h_{fz} \approx \frac{\Delta i_y}{\Delta i_x} \text{ where } v_{yz} = 0 \quad \text{(Eqn. 8-3c)}$$

$$h_{oz} \approx \frac{\Delta v_{xz}}{\Delta v_{yz}} \text{ where } i_x = 0 \quad \text{(Eqn. 8-3d)}$$

The "$\Delta$" indicates a slight change in the corresponding value about the BJT's quiescent point. The smaller the value of $\Delta$, the more precise the h-parameter value. For our analysis, we accept that "small change" means staying within ±10% of the quiescent point collector current.

Our BJT hybrid model has three limitations.

(1) Its h-parameters vary with collector current and temperature.

(2) Its h-parameters suffer from product variability, much as we saw with β.

(3) It does not address frequency effects.

Despite these limitations, results are generally accurate enough to guide us to a beginning design. Later, we can perfect the design by adjusting component values during simulation and breadboarding.

We can configure BJT amplifiers in three ways, depending on the application. These are *common emitter (CE)*, *common base (CB)*, and *common collector (CC, also called emitter follower)*. See the figure below.

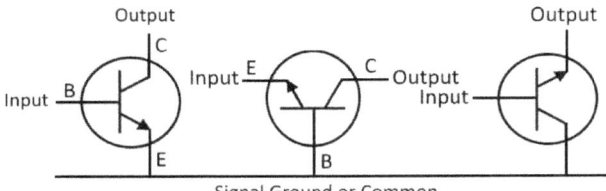

The "common" terms refer to which BJT element connects to the power supply common (the negative supply for NPN BJTs or the positive supply for PNP BJTs).

## Determining Hybrid Parameters

Hybrid parameters differ in value for the three configurations. In the hybrid model above, we use x, y, and z as generalized designations for the BJT's emitter, base, and collector. Below, we conduct our analysis using generalized h-parameters and find these amplifier *performance characteristics*: voltage gain $A_v$, current gain $A_i$, input resistance $R_{in}$, and output resistance $R_{out}$. When we later choose an amplifier configuration for a specific application, we will make the following substitutions:

*CE Configuration*: x = b for the BJT base, y = c for the BJT collector, and z = e for the BJT emitter. So, we would have $h_{ie}$, $h_{re}$, $h_{fe}$, and $h_{oe}$.

*CB Configuration*: x = e for the BJT emitter, y = c for the BJT collector, and z = b for the BJT base. So, we would have $h_{ib}$, $h_{rb}$, $h_{fb}$, and $h_{ob}$.

*CC Configuration*: x = b for the BJT base, y = e for the BJT emitter, and z = c for the BJT collector. So, we would have $h_{ic}$, $h_{rc}$, $h_{fc}$, and $h_{oc}$.

BJT data sheets do not usually provide information sufficient to calculate h-parameters using Eqns. 8-3 above. Occasionally, however, they do provide data or charts for CE h-parameters. Shown below are CE h-parameter graphs for the 2N3904 NPN BJT.

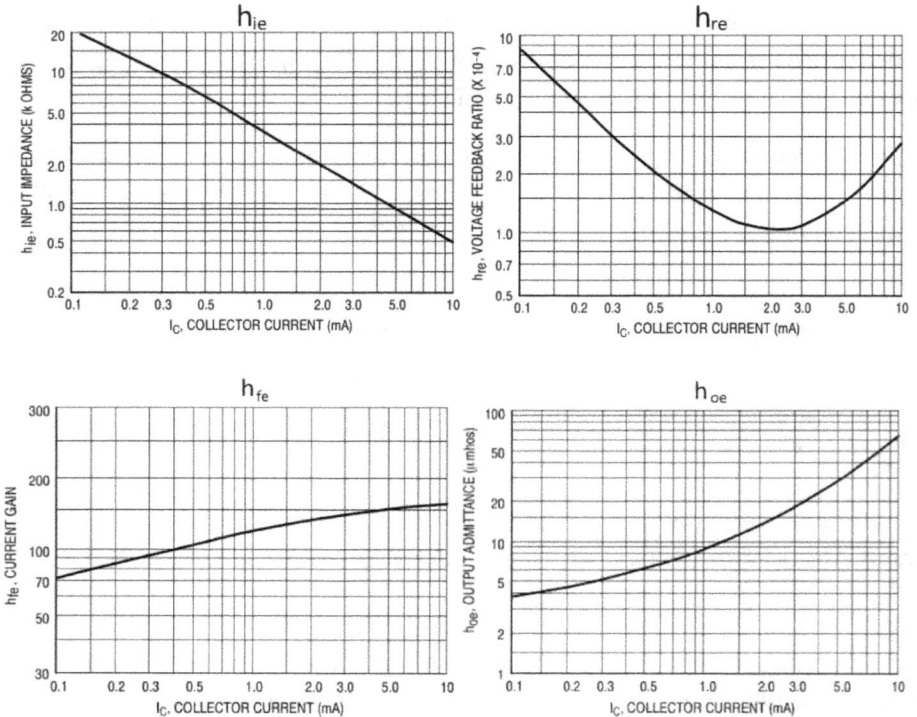

In each case, the graph shows a CE h-parameter plotted against the collector current $I_C$. The h-variation with $I_C$ emphasizes why we must limit current change to nearby (±10%) the quiescent point if we want reasonable results when using the model.

In executing a design, we first select a quiescent collector current; then, we find the corresponding CE h-parameters using these graphs. If we need CC or CB h-parameters, we use the formulas below.

Common Collector (Emitter Follower) Equivalents

$h_{ic} = h_{ie}$  (Eqn. 8-4a)

$h_{rc} = 1 - h_{re} \approx 1$  (Eqn. 8-4b)

$h_{fc} = -(h_{fe} + 1)$  (Eqn. 8-4c)

$h_{oc} = h_{oe}$  (Eqn. 8-4d)

Common Base Equivalents

$$h_{ib} = \frac{h_{ie}}{h_{fe} + 1}$$  (Eqn. 8-5a)

$$h_{rb} = \frac{h_{ie} \cdot h_{oe} - h_{re} \cdot (h_{fe} + 1)}{h_{fe} + 1}$$  (Eqn. 8-5b)

$$h_{fb} = -\frac{h_{fe}}{h_{fe} + 1}$$  (Eqn. 8-5c)

$$h_{ob} = \frac{h_{oe}}{h_{fe} + 1}$$  (Eqn. 8-5d)

If no h-parameter values are available for a small signal BJT, the values below give reasonable results in an early design phase.

CE

$h_{ie}$ = 1500 Ω
$h_{re}$ = $1.0 \cdot 10^{-4}$
$h_{fe}$ = 150
$h_{oe}$ = $18 \cdot 10^{-6}$ μmhos

Common Collector

$h_{ic}$ = 1500 Ω
$h_{rc}$ = 0.9999 ≈ 1.0
$h_{fc}$ = -151
$h_{oc}$ = $18 \cdot 10^{-6}$ μmhos

Common Base

$h_{ib}$ = 10 Ω
$h_{rb}$ = $7.9 \cdot 10^{-5}$
$h_{fb}$ = 0.9934
$h_{ob}$ = $1.19 \cdot 10^{-7}$ μmho

## Generalized h-Parameters Analysis

Our generalized h-parameter analysis of performance characteristics starts with the generalized hybrid BJT model like the one below.

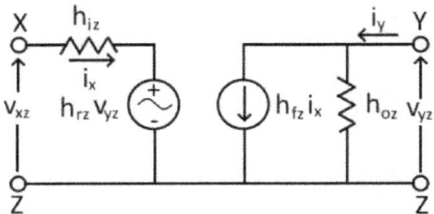

We then place the hybrid model into a basic amplifier circuit and calculate the amplifier's performance characteristics.

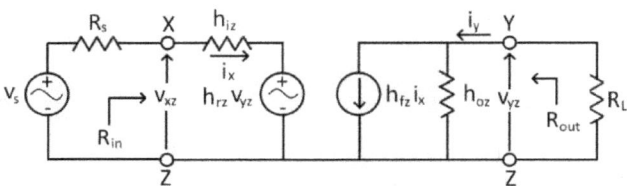

Note: We include the output resistance of the input source $R_s$ because it affects $R_{out}$. Likewise, we include load resistance $R_L$ because it affects both voltage gain $A_v$ and input resistance $R_{in}$.

We now proceed to derive the performance characteristics.

### Derive Voltage Gain $A_v$

First, we derive voltage gain $A_v = \dfrac{v_{yz}}{v_{xz}}$ for the basic BJT amplifier circuit below.

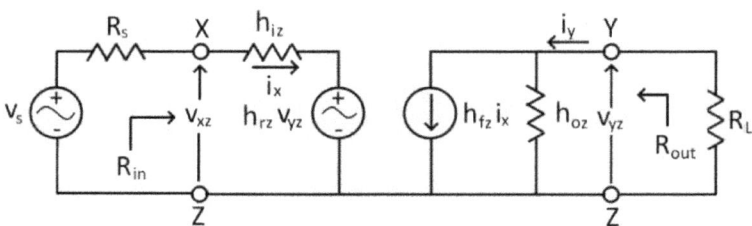

Recalling the generalized h-parameter equations, we have

$$v_{xz} = h_{iz} \cdot i_x + h_{rz} \cdot v_{yz} \quad \text{(Eqn. 8-6a)}$$

$$i_y = h_{fz} \cdot i_x + h_{oz} \cdot v_{yz} \quad \text{(Eqn. 8-6b)}$$

From Eqn. 8-6a,

$$i_x = \frac{v_{xz} - h_{rz} \cdot v_{yz}}{(h_{iz})} \quad \text{(Eqn. 8-7a)}$$

Given that the voltage across resistor $R_L$ is $v_{yz}$, then,

$$i_y = -\frac{v_{yz}}{R_L} \quad \text{(Eqn. 8-7b)}$$

When we equate Eqns. 8-6b and 8-7b and substitute in Eqn. 8-7a, we find that,

$$-\frac{v_{yz}}{R_L} = v_{yz} \cdot h_{oz} + h_{fz} \cdot \frac{v_{xz} - h_{rz} \cdot v_{yz}}{(h_{iz})} \quad \text{(Eqn. 8-7c)}$$

In Eqn. 8-7c, the only unknowns are $v_{yz}$ and $v_{xz}$ so that we can solve for their ratio. After algebraic manipulation and simplifying, we arrive at $A_v$.

$$A_v = \frac{v_{yz}}{v_{xz}} = \frac{-h_{fz} \cdot R_L}{(h_{iz}) \cdot (1 + h_{oz} \cdot R_L) - h_{fz} \cdot h_{rz} \cdot R_L} \quad \text{(Eqn. 8-8)}$$

Derive Current Gain $A_i$

We derive the current gain $A_i = i_y / i_x$ starting with Eqn. 8-6b,

$$i_y = h_{fz} \cdot i_x + h_{oz} \cdot v_{yz} \quad \text{(Eqn. 8-9a)}$$

Rearranging Eqn. 8-7b, we have,

$$v_{yz} = -i_y \cdot R_L \quad \text{(Eqn. 8-9b)}$$

When we substitute this equation into Eqn. 8-9a, this gives us,

$$i_y = h_{fz} \cdot i_x - h_{oz} \cdot i_y \cdot R_L \quad \text{(Eqn. 8-9c)}$$

Rearranging and taking the ratio of $i_y$ to $I_x$, we obtain our result,

$$A_i = \frac{h_{fz}}{(1 + h_{oz} \cdot R_L)} \quad \text{(Eqn. 8-10)}$$

Derive Input Resistance $R_{in}$

Next, we derive the input resistance $R_i$, which is $v_{xz} / i_x$. From Eqn. 8-6a, we have,

$$i_x = \frac{v_{xz} - h_{rz} \cdot v_{yz}}{h_{iz}} \quad \text{(Eqn. 8-11a)}$$

Combining Eqns. 8-6b and 8-5b, we find that,

$$i_y = -\frac{v_{yz}}{R_L} = v_{yz} \cdot h_{oz} + h_{fz} \cdot i_x \quad \text{(Eqn. 8-11b)}$$

Solving for $v_{yz}$,

$$v_{yz} = \frac{-h_{fz} \cdot i_x}{\left(h_{oz} + \frac{1}{R_L}\right)} \quad \text{(Eqn. 8-11c)}$$

When we substitute Eqn. 8-10c into Eqn. 8-10a and multiply by $h_{iz}$, we find that,

$$h_{iz} \cdot i_x = v_{xy} - h_{rz} \cdot \frac{-h_{fz} \cdot i_x}{\left(h_{oz} + \frac{1}{R_L}\right)} \quad \text{(Eqn. 8-11d)}$$

We now have an equation in the two unknowns, $v_{xy}$, and $i_x$, so we can solve for their ratio $v_{xz} / i_x$, which is the desired input resistance $R_{in}$.

$$R_{in} = h_{iz} - \frac{h_{rz} \cdot h_{fz}}{\left(h_{oz} + \frac{1}{R_L}\right)} \quad \text{(Eqn. 8-12)}$$

### Derive Output Resistance $R_{out}$

To derive the output resistance, we use the modified circuit below.

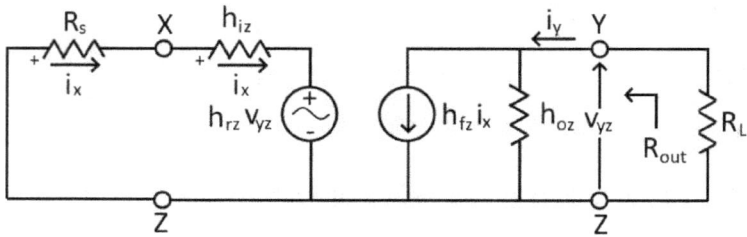

Note: For this derivation, we consider the signal source voltage $V_s = 0$ with only its internal resistance $R_s$ remaining.

Writing the loop equations for the input circuit, we have,

$$i_x \cdot R_s + i_x \cdot h_{iz} + h_{rz} \cdot v_{yz} = 0$$

Solving for $i_x$,

$$i_x = -\frac{h_{rz} \cdot v_{yz}}{R_s + h_{iz}} \quad \text{(Eqn. 8-13a)}$$

Recalling Eqn. 8-6b,

$$i_y = h_{fz} \cdot i_x + h_{oz} \cdot v_{yz} \quad \text{(Eqn. 8-13b)}$$

Replacing $i_x$ with Eqn. 8-13a,

$$i_y = h_{fz} \cdot -\frac{h_{rz} \cdot v_{yz}}{R_s + h_{iz}} + h_{oz} \cdot v_{yz} \quad \text{(Eqn. 8-13c)}$$

Solving for the ratio $v_{vz} / i_y$, which is $R_{out}$, we obtain the desired result,

$$R_{out} = \frac{1}{h_{oz} - \dfrac{h_{fz} \cdot h_{rz}}{R_s + h_{iz}}} \quad \text{(Eqn. 8-14)}$$

## Actual Amplifier Procedure

To find the performance characteristics of a CE, CC, or CB amplifier, we start with the generalized amplifier circuit below.

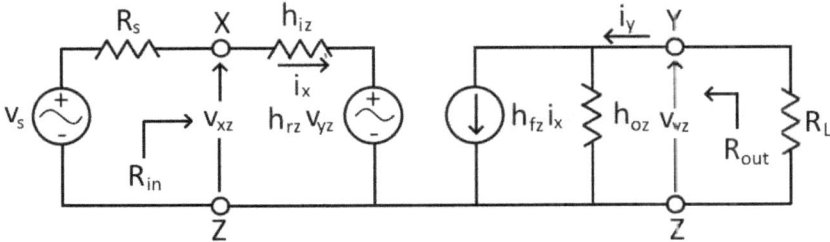

To adapt this generalized circuit to a CE, CC, or CB amplifier, we replace terminals X, Y, and Z with their respective terminal designations: B (base), C (collector), and E (emitter) for a CE amplifier; B, E, and C, for a CC amplifier; and E, C, and B, for a CB amplifier. Also, we replace the lower-case "z" designation of the h-parameter subscripts with a letter to designate the amplifier type: "e" for a CE amplifier, "c" for a CC amplifier, and "b" for a CB amplifier.

For example, a CE model amplifier would look like this,

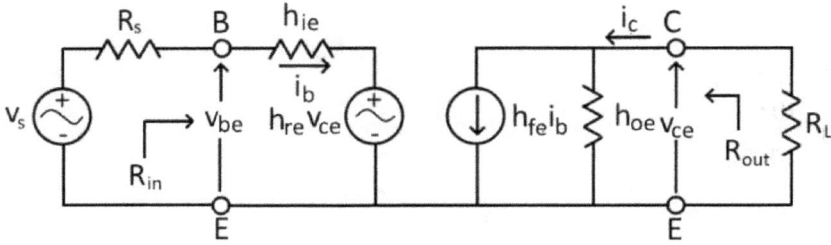

Next, we use the procedure in the section "Determining Hybrid Parameters" to obtain the h-parameter values for the amplifier type, whether CE, CC, or CB.

After obtaining h-parameters, we use the formulas below (with appropriate letter substitutions) to calculate the CE, CC, or CB amplifier's performance characteristics.

$$A_v = \frac{-h_{fz} \cdot R_L}{(h_{iz}) \cdot (1 + h_{oz} \cdot R_L) - h_{fz} \cdot h_{rz} \cdot R_L} \quad \text{(Eqn. 8-15a)}$$

$$A_i = \frac{h_{fz}}{(1 + h_{oz} \cdot R_L)} \quad \text{(Eqn. 8-15b)}$$

$$R_{in} = h_{iz} - \frac{h_{rz} \cdot h_{fz}}{\left(h_{oz} + \frac{1}{R_L}\right)} \quad \text{(Eqn. 8-15c)}$$

$$R_{out} = \frac{1}{h_{oz} - \frac{h_{fz} \cdot h_{rz}}{R_s + h_{iz}}} \quad \text{(Eqn. 8-15d)}$$

In the following chapters, we apply this procedure to CE, CC, and CB amplifier circuits.

# Chapter 9 – Common Emitter Amplifier

In the previous chapter, we derived amplifier performance characteristic equations for $A_v$, $A_i$, $R_{in}$, and $R_{out}$ using a generalized hybrid model. In this and the following chapters, we will apply these equations to design *single-ended BJT amplifiers.* We devote this chapter to the CE amplifier, which has these performance characteristics.

1. High voltage gain with inversion
2. High current gain with no inversion
3. Low input resistance
4. Low output resistance

## Bypassed Emitter Resistor

We will utilize the basic BJT circuit from Chapter 6 (circuit A below) and configure it as a CE amplifier (circuit B below).

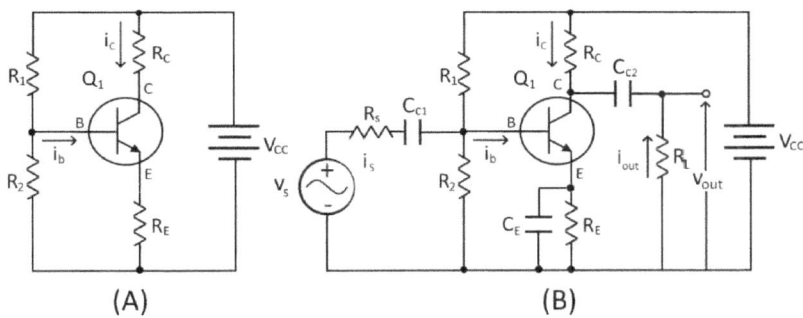

(A)                                                       (B)

In both, resistors $R_1$ and $R_2$ combine with emitter resistor $R_E$ to provide base bias and a stable quiescent collector current $I_c$.

In addition to the bias resistors, the CE amplifier consists of (1) a small signal source $V_S$ with an internal resistance $R_S$, (2) a load resistor $R_L$, (3) an emitter bypass capacitor $C_E$, and (4) coupling capacitors $C_{C1}$ and $C_{C2}$. We assume that capacitors $C_E$, $C_{C1}$, and $C_{C2}$ are short circuits at small signal frequencies. Bypass capacitor $C_E$ short-circuits AC signals on the emitter to the power supply common. $C_{C1}$ isolates the signal generator's DC path to common from the BJT base bias circuit. Otherwise, the signal generator would upset the base bias. Resistor $R_L$ functions as the load of the following stage's input circuit. $C_{C2}$ DC isolates the DC voltage on the collector from the following stage's input circuit. Without $C_{C2}$, resistor $R_L$ could adversely affect $Q_1$'s collector voltage.

In the small signal version below, we have replaced the BJT with its CE hybrid model.

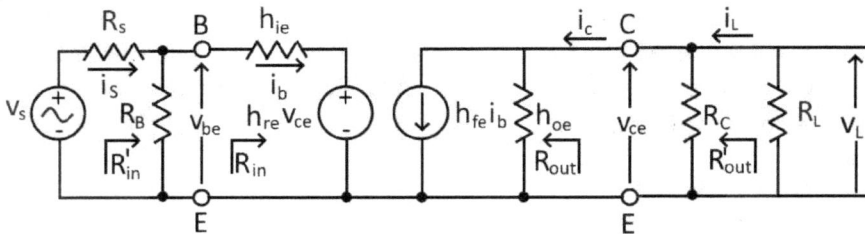

Note that the voltage supply $V_{CC}$ is a short circuit at signal frequencies. That being the case, we connected $R_B$ (the parallel combination of bias resistors $R_1$ and $R_2$) and $R_C$ to the power supply common and the emitter.

## Input Resistance – $R_{in}$

From the generalized model Eqn. 8-15c, we find that

$$R_{in} = h_{ie} - \frac{h_{re} \cdot h_{fe}}{\left(h_{oe} + \frac{1}{R'_L}\right)} \quad \text{where } R'_L = R_C || R_L. \text{ (Eqn. 9-1a)}$$

When we include the bias resistors,

$$R'_{in} = R_{in} || R_B \quad \text{(Eqn. 9-1b)}$$

Note: Symbols with a prime " ' " or a double prime " " " indicate their values contain components <u>external</u> to the BJT model. Without the prime, they represent values internal to the BJT itself.

With the emitter bypassed, $R'_{in} = R_B \,||\, h_{ie}$. Since $h_{ie}$ is at most a few thousand ohms, we know that $R'_{in}$ will be smaller than $R_B$. This accounts for the CE amplifier's low input resistance.

## Voltage Gain- Av

Using the voltage divider law, we arrive at,

$$v_{be} = v_s \cdot \frac{R'_{in}}{R'_{in} + R_s} \quad \text{(Eqn. 9-2)}$$

From Eqn. 8-15a, we find that,

$$A_v = \frac{v_{ce}}{v_{be}} = \frac{-h_{fe} \cdot R'_L}{(h_{ie}) \cdot (1 + h_{oe} \cdot R'_L) - h_{fe} \cdot h_{re} \cdot R'_L} \text{where } R'_L$$
$$= R_C || R_L \quad \text{(Eqn. 9-3)}$$

If we neglect the effects of $h_{re}$ and $h_{oe}$, the voltage gain is,

$$A_v = \frac{-h_{fe} \cdot R'_L}{h_{ie}}$$

Since $R'_L$ and $h_{ie}$ are of the same order of magnitude, then $A_v \approx h_{fe}$. This is consistent with the statement above that the CE amplifier voltage gain is high. It also means that its voltage gain is highly dependent on $h_{fe}$ and, therefore, widely variable. We will come back to this later and explore the way around it.

Combining Eqns. 9-2 and 9—3, we find that,

$$A'_v = \frac{v_{ce}}{v_S} = \frac{v_{out}}{v_S}$$
$$= \frac{R'_{in}}{R'_{in} + R_S}$$
$$\cdot \frac{-h_{fe} \cdot R'_L}{(h_{ie}) \cdot (1 + h_{oe} \cdot R'_L) - h_{fe} \cdot h_{re} \cdot R'_L} \quad \text{(Eqn. 9-4a)}$$

If we neglect the effects of $h_{re}$ and $h_{oe}$, the voltage gain is then,

$$A'_v = \frac{R''_{in}}{R''_{in} + R_S} \cdot \frac{-h_{fe} \cdot R'_L}{h_{ie}} \quad \text{where } R''_{in} = R_B || h_{ie} \quad \text{(Eqn. 9-4b)}$$

Keeping $R_S << R'_{in}$ lessens gain loss introduced by the first factor in the equation.

## Current Gain- $A_i$

Moving now to the current gain $A_i$, we will use the circuit below.

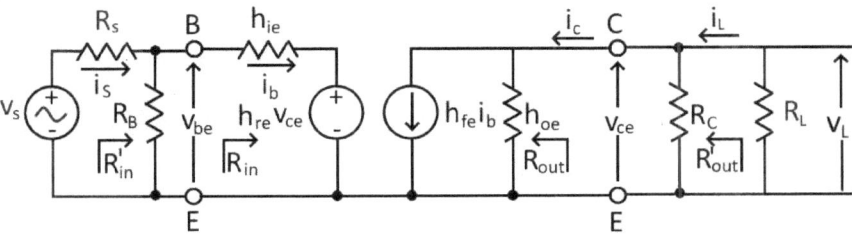

Given the input current $i_S$, then by the Current Divider Law,

$$i_b = \frac{R_B}{R_B + R_{in}} \cdot i_s$$

From the generalized Eqn. 8-15b, we obtain,

$$\frac{i_c}{i_s} = \frac{R_B}{R_B + R_{in}} \cdot \frac{h_{fe}}{(1 + h_{oe} \cdot R'_L)}$$

To find the output current $i_L$ (that flows in load resistor $R_L$), we use the Current Divider Law again.

$$A'_i = \frac{i_L}{i_s} = \frac{R_B}{R_B + R_{in}} \cdot \frac{h_{fe}}{(1 + h_{oe} \cdot R'_L)} \cdot \frac{R_C}{R_C + R_L} \quad \text{(Eqn.9-5a)}$$

In practice, $h_{oe}$ has much less effect on current gain than $R_B$ and $R_C$. Neglecting $h_{oe}$, we find that,

$$A'_i = \frac{i_L}{i_s} = \frac{R_B}{R_B + R_{in}} \cdot h_{fe} \cdot \frac{R_C}{R_C + R_L} \quad \text{(Eqn.9-5b)}$$

Keeping $R_B$ and $R_C$ as large as possible keeps current gain high. Like voltage gain, however, the current gain is highly dependent on $h_{fe}$, which is highly variable.

## Output Resistance – $R_{out}$

Next, we use Eqn. 8-15d to obtain the output resistance.

$$R_{out} = \frac{1}{h_{oe} - \dfrac{h_{fe} \cdot h_{re}}{R'_s + h_{ie}}} \quad \text{(Eqn. 9-6)}$$

The amplifier's output resistance $R'_{out}$ includes collector resistor $R_C$, as shown below.

$$R'_{out} = \frac{1}{h_{oe} - \dfrac{h_{fe} \cdot h_{re}}{R'_s + h_{ie}}} \mathbin{||} R_c \quad \text{(Eqn. 9-7)}$$

In practice, $R_C \ll \frac{1}{h_{oe}}$, so $R_C$ dominates the value of output resistance, and since $R_C$ is in the few 1000s of $\Omega$ or less, output resistance is typically low.

## CE Example

To verify these results, we used the CE amplifier below.

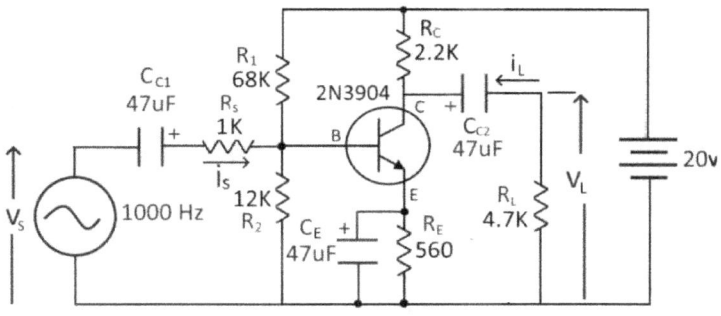

For the breadboard test, we planned to use a signal generator with an output resistance of only 50 Ω. Because this was too low to verify the effect of Rs properly, we added the 1000 Ω resistor shown, making Rs = 1050 Ω. The 47 μF emitter bypass capacitor CE bypassed the 1000 Hz small signal to the power supply common. The 47 μF coupling capacitor CC1 was necessary to isolate the signal generator output from BJT's bias resistors. Otherwise, the signal generator would provide a DC path from base to common and upset the BJT's base bias. Lastly, we used a 47 μF capacitor CC2 to couple load RL to the BJT collector circuit. Without it, RL would provide an undesirable DC path to common.

Recall that the quiescent collector current for the basic CE amplifier was 3.7 ma. From the h-parameter charts in Chapter 8, we found that,

$h_{ie}$ = 1200 Ω
$h_{re}$ = 1.4 · 10⁻⁴
$h_{fe}$ = 150
$h_{oe}$ = 22 · 10⁻⁶ mhos

With this information, we performed the following calculations:

$$R_B = \frac{68 \cdot 12}{68 + 12} = 10.2K$$

$$R'_L = \frac{2.2 \cdot 4.7}{2.2 + 4.7} = 1.5K$$

$$R_{in} = 1200 - \frac{1.4 \cdot 10^{-4} \cdot 150}{\left(22 \cdot 10^{-6} + \frac{1}{1500}\right)} = 1170 \ \Omega$$

$$R'_{in} = R_{in}||R_B = \frac{10200 \cdot 1170}{10200 + 1170} = 1050 \ \Omega$$

Then, using Eqn. 9-4, we found,

$$A'_v = \frac{v_{ce}}{v_s} = \frac{v_{out}}{v_s}$$

$$= \frac{1050}{1000 + 1050}$$

$$\cdot \frac{-150 \cdot 1500}{(1200) \cdot (1 + 22 \cdot 10^{-6} \cdot 1500) - 150 \cdot 1.4 \cdot 10^{-4} \cdot 1500}$$

$$A'_v = -93$$

Next, we calculated output resistance R$_{out}$.

$$R_{out} = \frac{1}{22 \cdot 10^{-6} - \frac{150 \cdot 150 \cdot 1.4 \cdot 10^{-4}}{910 + 1200}} = 83,000 \ \Omega$$

Taking R$_C$ into account, we found that,

$$R'_{out} = \frac{83,000 \cdot 2200}{83,000 + 2200} = 2143 \ \Omega$$

To calculate A'$_i$, we used Eqns. 9-5.

$$A'_i = \frac{i_L}{i_s} = \frac{R_B}{R_B + R_{in}} \cdot \frac{h_{fe}}{(1 + h_{oe} \cdot R'_L)} \cdot \frac{R_C}{R_C + R_L}$$

$$A'_i = \frac{i_L}{i_s} = \frac{10200}{10200 + 1170} \cdot \frac{150}{(1 + 22 \cdot 10^{-6} \cdot 1500)} \cdot \frac{2143}{2143 + 4700} = 42$$

For the test, we used a 2N3904 with a $\beta \approx 150$. The results were as follows:

|  | Expected | Simulation | Breadboard |
|---|---|---|---|
| A'$_v$ | -95 | -97 | -97 |
| A'$_i$ | 42 | 43 | 40 |
| R'$_{in}$ | 1050 $\Omega$ | 1079 $\Omega$ | 1200 $\Omega$ |
| R'$_{out}$ | 2143 $\Omega$ | 2190 $\Omega$ | 2100 $\Omega$ |

The results were good. Keep in mind, however, that we used a BJT with $h_{fe}$ = 150 in the calculation, simulation, and breadboard. In the real world, production values of $h_{fe}$ and the other h-parameters would have varied, making the results much less satisfactory.

## Unbypassed Emitter Resistor

Bypassing $R_E$ with capacitor $C_E$ in the circuit above did not affect DC stability. It did, however, eliminate AC negative feedback and AC stability that would have been desirable. To study the effects of $R_E$ on AC stability, we used the hybrid model circuit below.

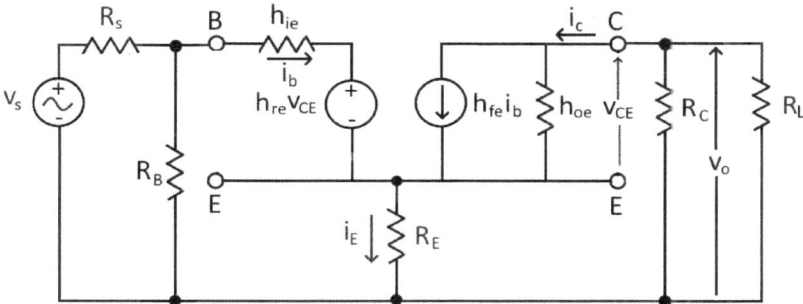

To simplify our analysis, we made two assumptions:

1. The portion of $i_c$ flowing through $h_{oe}$ is negligible compared to the current source $h_{fe} \cdot i_b$, so we will neglect the effect of $h_{oe}$.

2. Input voltage $v_{be}$ was larger than $h_{re} \cdot v_{ce}$, so we will neglect the effect of $h_{re}$.

With these assumptions made, our model circuit was reduced to,

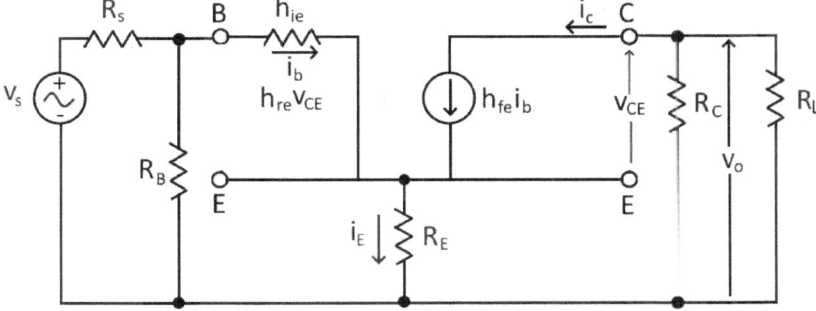

The first thing we observed was that $i_E = i_b + i_c = i_b + h_{fe} \cdot i_b = (h_{fe} + 1) \cdot i_b$. From this, we saw that the voltage drop across $R_E$ equaled $(h_{fe} + 1) \cdot R_E \cdot i_b$. This suggested that the voltage drop from the emitter to common is the same as if we were to replace resistor $R_E$ with a resistance equal to $(h_{fe} + 1) \cdot R_E$. With this change, the model circuit became,

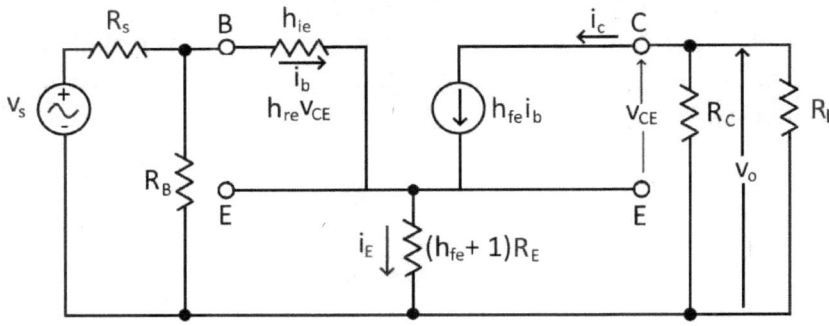

We were now able to calculate the CE amplifier's performance characteristics.

Voltage Gain- A<sub>v</sub>

Voltage Gain- A$_v$

To find voltage gain, we first observed that,

$$i_b = \frac{v_B}{h_{ie} + (h_{fe} + 1) \cdot R_E}$$

Knowing that $v_o = -(R_c||R_L) \cdot i_c$ and $i_c = h_{fe} \cdot i_b$, we found that,

$$v_o = -(R_c||R_L) \cdot i_c = -(R_c||R_L) \cdot h_{fe} \cdot i_b = -R'_L \cdot h_{fe} \cdot \frac{v_B}{h_{ie} + (h_{fe} + 1) \cdot R_E}$$

where $R'_L = R_c||R_L = \frac{R_C \cdot R_L}{R_C + R_L}$.

The voltage gain then is,

$$A_v = \frac{v_o}{v_B} = \frac{-R'_L \cdot h_{fe}}{h_{ie} + (h_{fe} + 1) \cdot R_E} \qquad \text{(Eqn. 9-8a)}$$

Taking the source resistance R$_s$ into account, we found that,

$$A'_v = \frac{v_o}{v_s} = \frac{R'_B}{R_s + R'_B} \cdot \frac{-R'_L \cdot h_{fe}}{h_{ie} + (h_{fe} + 1) \cdot R_E} \text{ where } R'_B$$
$$= [h_{ie} + (h_{fe} + 1) \cdot R_E] \,\|\, R_B \quad \text{(Eqn. 9-8b)}$$

As a check, we made $R_E = 0$ and Eqn.9-8b reduced to Eqn. 9-4b.

If we assume (1) $h_{fe} > 100$ then $(h_{fe} + 1) \cdot R_E \gg h_{ie}$, and (2) $R_s \ll R'_B$. this useful approximation resulted,

$$A'_v \approx -\frac{R'_L}{R_E} \quad \text{(Eqn. 9-9)}$$

Since a production range of $100 < \beta < 300$ easily meets assumption (1) above, we concluded that the AC negative feedback provided by unbypassed resistor $R_E$ stabilized voltage gain at the ratio of load resistance $R'_L$ to emitter resistor $R_E$. By removing the effect of $\beta$ variability, we made it possible to trade voltage gain for improved AC stability!

## Input Resistance – $R_{in}$
We further observed that,

$$R_{in} = R_B \,\|\, (h_{ie} + 1)R_E \quad \text{(Eqn. 9-10)}$$

Since $(h_{ie} + 1) \cdot R_E$ is typically large, $R_B$ controls the value of $R_{in}$. So, we can now use $R_B$ to calculate a specified input resistance.

## Output Resistance - $R_{out}$
$$R_{out} = \left(h_{oe} + (h_{fe} + 1) \cdot R_E\right) \,\|\, R_c$$

Also, since typically $(h_{fe}+1) \cdot R_E \gg R_c$, we can simplify the equation to,

$$R_{out} = R_c \quad \text{(Eqn. 9-11)}$$

## Current Gain – $A_i$
Moving now to the current gain, we used the figure below.

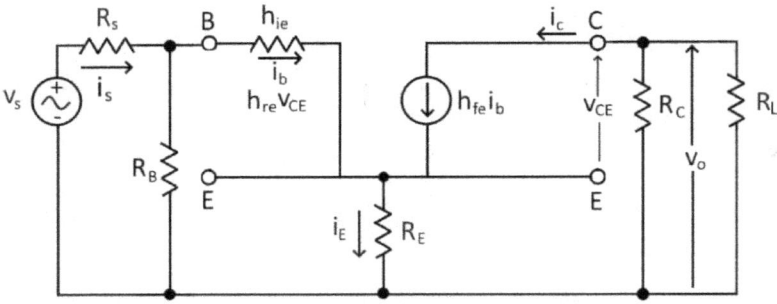

We began by noting that,

$$i_o = \cdot\, i_b$$

and therefore,

$$A_i = \frac{i_o}{i_B} = h_{fe}$$

Using the Current Divider Law, we calculated base current $i_b$ in terms of $i_s$ as,

$$i_b = \frac{R_B}{R_{in} + R_B} \cdot i_s$$

Using the Current Divider Law on the output, we found that,

$$i_o = \frac{R_c}{R_L + R_c} \cdot i_c = \frac{R_c}{R_L + R_c} \cdot h_{fe} \cdot i_b = \frac{R_c}{R_C + R_L} \cdot \frac{h_{fe} \cdot R'_s}{R_{in} + R'_s} \cdot i_s$$

and finally,

$$A'_i = \frac{i_o}{i_s} = \frac{R_B}{R_{in} + R_B} \cdot h_{fe} \cdot \frac{R_c}{R_C + R_L} \qquad \text{(Eqn. 9-11)}$$

To check, when we made $h_{oe} = 0$ and $h_{re} = 0$ in Eqn. 9-5b and $R_E = 0$ in Eqn. 9-11, then the two equations for $A_i$ were the same.

## CE Example

To verify our calculations, we used our CE amplifier with the emitter bypass capacitor removed.

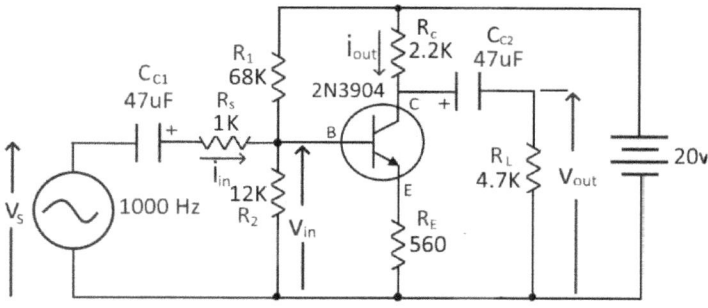

The calculated voltage gain was,

$$A_v = \frac{v_{out}}{v_{in}} \approx -\frac{R_c \| R_K}{R_E} = -\frac{2200 \| 4700}{560} = -\frac{1500}{560} = 2.7$$

In the emitter unbypassed configuration, the design focuses on voltage gain, so we limited our test to $A_v$. The test results were,

|       | Expected | Simulation | Breadboard |
|-------|----------|------------|------------|
| $A_v$ | -2.7     | -2.6       | -2.6       |

The results were good. Moreover, we tried BJTs with different $h_{fe}$ values. There was no meaningful change in voltage gain!

Suppose we want to increase the gain without recalculating the bias resistors. We can accomplish this by splitting the emitter resistor and bypassing by whatever portion gives the desired $R'_L / R_E$ ratio. For example, suppose we wanted,

$$A_v = \frac{v_{out}}{v_{in}} = 10$$

We would need an unbypassed resistance,

$$R_{E1} = \frac{1500}{10} = 150 \ \Omega$$

and a bypassed resistance of

$$R_{E2} = 560 - 150 = 410 \ \Omega$$

The nearest 5% value for $R_{E2}$ is 390 Ω. The amplifier circuit then becomes,

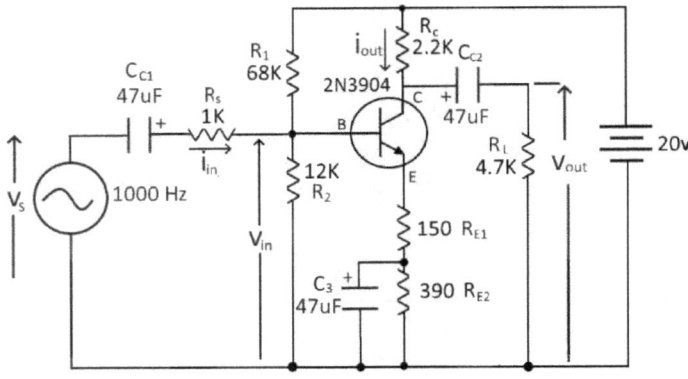

When tested, the results were,

|  | Expected | Simulation | Breadboard |
|---|---|---|---|
| $A_v$ | -10 | -9.4 | -9.5 |

Using 1% resistors in the breadboard circuit, we could have tweaked the value of $R_{E1}$ to give $A_v$ = 10. Practically speaking, however, we rarely need such precise gains when working with BJT amplifiers. Keep in mind that by decreasing the unbypassed emitter resistance, we also decrease AC stability. However, we tried BJTs with different $h_{fe}$ values with a negligible change in voltage gain!

The CE amplifier is the workhorse of BJT amplifiers when high voltage and/or current gain is paramount. When AC stability is essential, the unbypassed $R_E$ version is best. To obtain a higher voltage gain, we can use multiple unbypassed emitter stages.

Other situations dictate that we consider the CC and CB amplifier configurations. For instance, when matching input and output resistance is our design focus, we would use the CC configuration. Or, to build a high-frequency amplifier, we would use the CB configuration. We consider these in the following chapters.

# Chapter 10 – Common Collector Amplifier

The CC amplifier or emitter follower, as it is also known, has these performance characteristics:

1. Voltage gain of 1 with no inversion
2. High current gain with no inversion
3. High input resistance
4. Exceptionally low output resistance

This makes it particularly suitable as a load-matching device, i.e., designing the CC amplifier's input resistance to match a source device's requirement.

In an NPN CC amplifier, the collector connects to the power supply plus. In a PNP CC amplifier, the collector connects to the power supply negative. In both cases, the power supply is a short circuit at signal frequencies, which, in effect, connects the collector to the power supply's common as required for CC amplifiers.

In a CC amplifier, we apply the input signal $v_{in}$ at the BJT base and take the output signal $v_{out}$ from the emitter. See the figure below.

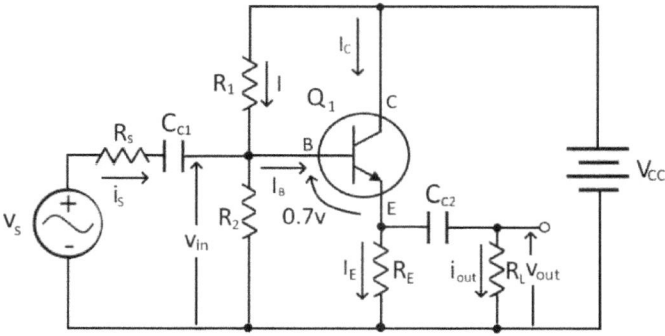

To do a small signal analysis of the CC amplifier, we replaced the BJT with the CC hybrid model.

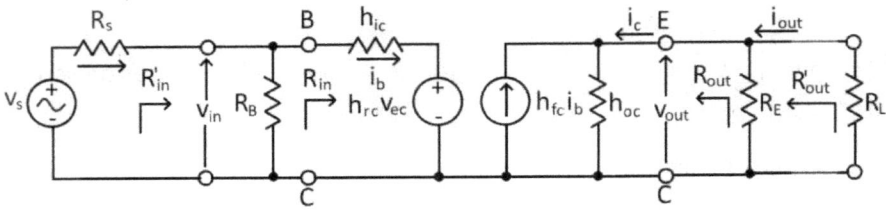

The amplifier's performance characteristics are those of Eqns. 8-15 with CC h-parameters replacing the generalized h-parameters.

$$A_v = \frac{v_o}{v_{in}} = \frac{-h_{fc} \cdot R_E}{(h_{ic} + R_s) \cdot (1 + h_{oc} \cdot R_E) - h_{fc} \cdot h_{rc} \cdot R_E} \quad \text{(Eqn. 10-1a)}$$

$$A_i = \frac{i_c}{i_b} = \frac{-h_{fc}}{(1 + h_{oc} \cdot R_E)} \quad \text{(Eqn. 10-1b)}$$

$$R_{in} = h_{ic} - \frac{h_{rc} \cdot h_{fc}}{\left(h_{oc} + \frac{1}{R_E}\right)} \quad \text{(Eqn. 10-1c)}$$

$$R_{out} = \frac{1}{h_{oc} - \frac{h_{fc} \cdot h_{rc}}{R_s + h_{ic}}} \quad \text{(Eqn. 10-1d)}$$

## Voltage Gain- A$_v$

First, we find voltage gain,

$$A_v = \frac{-(h_{fc}) \cdot R'_E}{(h_{ic}) \cdot (1 + h_{oc} \cdot R'_E) - h_{fc} \cdot h_{rc} \cdot R'_E)}$$
$$= \frac{1}{1 - \frac{(h_{ic}) \cdot (1 + h_{oe} \cdot R_{iE})}{h_{fc} \cdot R'_E}} \quad \text{(Eqn. 10-2)}$$

where $R'_E = R_E \parallel R_L$.

For $h_{oc} \cdot R'_E \ll 1$ and $h_{ic} \ll -h_{fc} \cdot R'_E$,

$$A_v \approx 1 \quad \text{(Eqn. 10-3a)}$$

When we take R$_s$ into effect, we reduce voltage gain by the factor $\frac{R'_{in}}{R'_{in} + R_S}$ where R'$_{in}$ = R$_B$ || R$_{in}$. As R$_{in}$ is much larger than R$_B$, the factor becomes $\frac{R_B}{R_B + R_S}$. .

Therefore, we find that,

$$A'_v = \frac{R_B}{R_B + R_S} \cdot A_v \quad \text{(Eqn. 10-3b)}$$

In a matching application, $R_B = R_S$ and $A'_v = 0.5$.

## Current Gain – $A_i$

Next, we find the current gain.

$$A_i = \frac{-h_{fc}}{(1 + h_{oc} \cdot R_E)} \quad \text{(Eqn. 10-4)}$$

For $h_{oe} \cdot R_E \ll 1$,

$$A_i \approx -h_{fc} \quad \text{(Eqn. 10-5a)}$$

The current gain is high and not inverted (recall that $h_{fc}$ = -$h_{fe}$). If $i_{out}$ is the output current, then we must multiply by the factor $\frac{R_E}{R_E+R_L}$. The current gain then becomes,

$$A'_i \approx -h_{fc} \cdot \frac{R_E}{R_E + R_L} \quad \text{(Eqn. 10-5b)}$$

## Input Resistance- $R_{in}$

Moving now to the input resistance $R_{in}$, we find that,

$$R_{in} = h_{ic} - \frac{h_{rc} \cdot h_{fc}}{\left(h_{oc} + \frac{1}{R_E}\right)} = h_{ic} - \frac{h_{fc} \cdot R_E}{(h_{oc} \cdot R_E + 1)} \quad \text{(Eqn. 10-6)}$$

For $h_{oc} \cdot R_E \ll 1$,

$$R_{in} \approx h_{ic} - h_{fc} \cdot R_E \quad \text{(Eqn. 10-7a)}$$

Taking $R_B$ into account, we find that,

$$R'_{in} \approx (h_{ic} - h_{fc} \cdot R_E) \parallel R_B \quad \text{(Eqn. 10-7b)}$$

As noted earlier, usually $R_{in} \gg R_B$, so we can assume that,

$$R'_{in} \approx R_B \quad \text{(Eqn. 10-7c)}$$

Given this fact, we can choose $R_B$ to match the input resistance $R_s$ requirement of the source device.

## Output Resistance — $R_{out}$

And lastly, we find $R_{out}$,

$$R_{out} = \cfrac{1}{h_{oc} - \cfrac{h_{fc} \cdot h_{rc}}{h_{ic}}} \qquad \text{(Eqn. 10-8)}$$

$$R_{out} = \frac{h_{ic}}{h_{ic}.h_{oc} - h_{fc}} \qquad \text{(Eqn. 10-9)}$$

For $h_{ic} \cdot h_{oc} << -h_{fc}$,

$$R_{out} \approx \frac{-h_{ic}}{h_{fc}} \qquad \text{(Eqn. 10-10a)}$$

With $h_{ic}$ on the order of 1000 to 2000 $\Omega$, $R_{out}$ is small, less than 20 $\Omega$. $R_{out}$ being this small means we can ignore the effect of $R_E$ and say that,

$$R'_{out} \approx \frac{-h_{ic}}{h_{fc}} \qquad \text{(Eqn. 10-10b)}$$

## CC Example

For our example, we used a CC amplifier to match the specified output resistance of a magnetic phono cartridge, 47 KΩ. We also assumed that the load resistance was 5 KΩ.

Recalling that the collector current is 2 ma, we used the charts in Chapter 8 to find these CE h-parameters for the 2N3904:

$h_{ie}$ = 2000 Ω
$h_{re}$ = 1.1 · $10^{-4}$
$h_{fe}$ = 140
$h_{oe}$ = 13 · $10^{-6}$ mhos

The CC h-parameters were then,

$h_{ic}$ = 2000 Ω
$h_{rc}$ = 1
$h_{fc}$ = -141
$h_{oc}$ = 9.22 · $10^{-8}$ mhos

As noted earlier, if we choose equal bias resistors, the DC output voltage $V_E$ will be $V_{cc}/2$. If $V_{cc}$ is 20 v, then $V_E$ will then be 10 volts. Knowing this, we chose the value of $R_E$. If $I_E \approx I_c = 2$ ma, then,

$$R_E = \frac{10}{2 \cdot 10^{-3}} = 5 \text{ K}\Omega$$

In this case, $R_{in}$ was,

$$R_{in} = 2000 - (-141 + 1) \cdot 5 \cdot 10^3 = 702 \text{ K}\Omega$$

Given this large value for $R_{in}$, we made $R_1$ and $R_2$ twice 47 KΩ. Using standard values of 100 KΩ gave $R_1 \parallel R_2 = 50$ KΩ, close enough to our target value of 47 KΩ.

The example circuit is below.

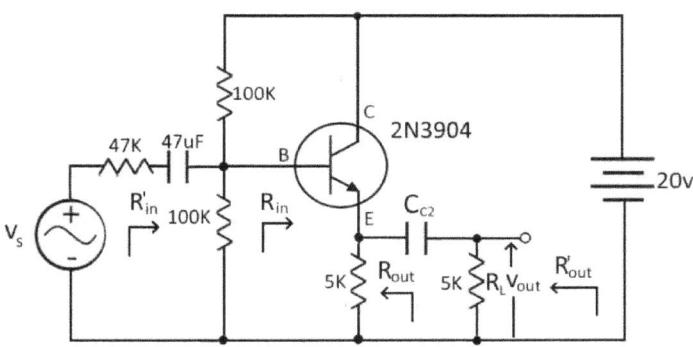

For our circuit, here are the performance characteristics.

$$R'_E = 5K \parallel 5K = 2.5K$$

$$A_v = \cfrac{1}{1 - \cfrac{(h_{ic}) \cdot (1 + h_{oe} \cdot R'_E)}{h_{fc} \cdot R_{\prime E}}} = \cfrac{1}{1 - \cfrac{(2000) \cdot (1 + 13 \cdot 10^{-6} \cdot 2500)}{-141 \cdot 2500}}$$
$$= 0.9941 \ (\approx 1 \text{ using Eqn 10-3})$$

$$A'_v \approx \frac{47}{47 + 50} \cdot 0.9941 = 0.4971$$

$$A_i = \frac{-h_{fc}}{(1 + h_{oc} \cdot R_E)} = \frac{-141}{(1 + 13 \cdot 10^{-6} \cdot 5000)} = 132 \ (\approx 141 \text{ using Eqn. 10-5})$$

$$A'_i = \frac{5}{5+5} = \frac{1}{2} \cdot 132 = 66$$

$$R_{in} = h_{ic} - \frac{h_{fc} \cdot R_E}{(h_{oc} \cdot R_E + 1)} = 2000 - \frac{-141 \cdot 5000}{(1 + 13 \cdot 10^{-6} \cdot 5000)}$$
$$= 662 \; K\Omega \; (\approx 707 \; K\Omega \text{ using Eqn. 10-7})$$

$$R'_{in} = \frac{662 \cdot 50}{662 + 50} = 46.5 \; K\Omega$$

This value brings $R'_{in}$ even closer to our 47 KΩ target value!

$$R_{out} = \frac{h_{ic}}{h_{ic}.h_{oc} - h_{fc}} = \frac{2000}{2000 \cdot 13 \cdot 10^{-6} - (-141)}$$
$$= 14\Omega \; (\approx 14\Omega \text{ using 10-10})$$

$$R'_{out} = \frac{14 \cdot 5000}{14 + 5000} \approx 14 \; \Omega$$

In each case, we included the approximately calculated value in parentheses. We would not have been far off to use these in the first place!

For our test CC amplifier, we measured these performance characteristics:

|           | Expected | Simulation | Breadboard |
|-----------|----------|------------|------------|
| $A_v$     | 1        | 0.995      | 1.0 (within the precision we could measure) |
| $A_i$     | 132      | 141        | 138        |
| $R'_{in}$ | 47 KΩ    | 47 KΩ      | 48.5 KΩ (used 5% 100 KΩ resistors) |
| $R'_{out}$| 14 Ω     | 55 Ω       | 43 Ω       |

Except for $R'_{out}$, the value agreement was good. As seen in Eqn. 10-10a, $R'_{out}$ depends on the value of $h_{ic}$, which was not precisely known in the simulation or breadboard BJT.

What we take away is that the output resistance will be low, easily less than 100 Ω and the input resistance will be close to the value of $R_B$. Both results suggest a CC stage works well as a load-matching device, as in a magnetic phono cartridge example. The CC's low output resistance finds application driving low-resistance devices like relays and speakers.

In the next chapter, we consider the last of the three BJT configurations, the CB amplifier.

# Chapter 11 – Common Base Amplifier

The CB amplifier has these performance characteristics:

1. High voltage gain with no inversion
2. Low current gain with inversion
3. Low input resistance
4. High output resistance

The CB amplifier configuration (see below) is the least often seen of the three BJT configurations.

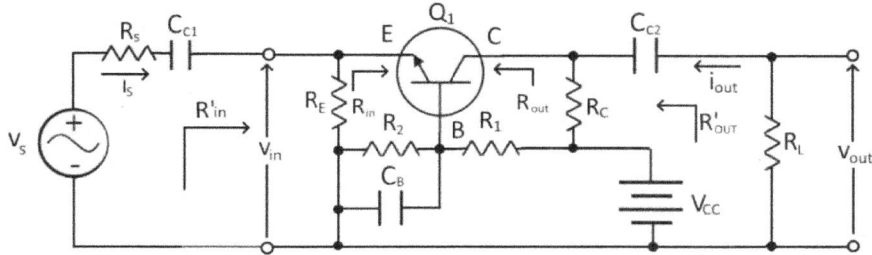

The biasing arrangement resembles its CE cousin with $R_1$, $R_2$, and, this time, $R_E$, setting the base bias. The base connects to common for small signals via capacitor $C_B$. For the CB amplifier, we apply the input signal to the emitter and take the output from the collector.

Note: It is possible to directly connect the base to the common if we use a split $\pm V_{cc}$ power supply.

Recall that the CC amplifier has high input resistance, exceptionally low output resistance, and a voltage gain of 1. The CB is just the opposite. It has extremely low input resistance, high output resistance, and high voltage gain. We use a CB amplifier when we need to couple to a low-output resistance device or circuit. Years ago, the author made an intercom in which the speakers served also as the microphone. The speaker output resistance was 8 Ω, which matched nicely with a CB amplifier's exceptionally low (~12 Ω) input resistance.

We also find CB amplifiers used as radio frequency amplifiers where tuned circuits have exceptionally low impedances. CB amplifiers have a better gain at high frequencies than CE or CC amplifiers. This results from the fact that the common-connected base acts as a shield between the collector and emitter. We will say more about this later.

Next, we use the hybrid model to compute CB performance characteristics.

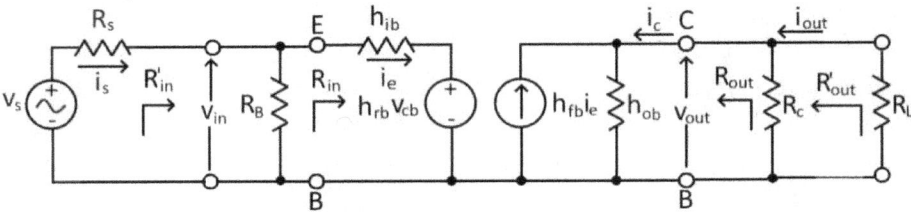

Using the generalized Eqns. 8-15 with CC h-parameters, we find that,

## Voltage Gain- A$_v$

$$A_v = \frac{v_{out}}{v_{in}} = \frac{-h_{fb} \cdot R_C}{(h_{ib}) \cdot (1 + h_{ob} \cdot R_C) - h_{fb} \cdot h_{rb} \cdot R_C} \quad \text{(Eqn. 11-1a)}$$

To find the voltage gain relative to v$_s$, we multiply by the factor $\frac{R_{in}}{R_{in}+R_s}$. Therefore,

$$A'_v = \frac{v_{out}}{v_s} = \frac{R_{in}}{R_{in} + R_s} \cdot A_v \quad \text{(Eqn.11-1b)}$$

R$_{in}$ is exceedingly small, usually 20 Ω or less. Unless R$_s$ is also small, the factor will drastically reduce voltage gain. The limited application when this is the case explains the rare appearance of the CB configuration.

## Input Resistance- R$_{in}$

$$R_{in} = h_{ib} - \frac{h_{rb} \cdot h_{fb}}{\left(h_{ob} + \frac{1}{R_C}\right)} \quad \text{(Eqn. 11-2a)}$$

Because R$_{in}$ is so small, paralleling it with R$_B$ has no effect. So, we find that,

$$R'_{in} = R_{in} \quad \text{(Eqn.11-2b)}$$

## Current Gain — A$_i$

$$A_i = \frac{i_c}{i_E} = \frac{h_{fb}}{(1 + h_{ob} \cdot R_C)} \quad \text{(Eqn. 11-3a)}$$

Since $h_{ob} \cdot R_c \ll 1$, the current gain is $h_{fb}$, which is ~-1. If we want the current gain with respect to $i_s$ and $i_{out}$, then we must multiply by these factors, $\frac{R_s}{R_{in}+R_s}$ equation and $\frac{R_{out}}{R_{out}+R_L}$. Therefore,

$$A'_i = \frac{i_{out}}{i_s} = \frac{R_B}{R_{in} + R_B} \cdot \frac{R_{out}}{R_{out} + R_L} \cdot A_i \quad \text{(Eqn. 11-3b)}$$

Since $R_{out} \gg R_L$, we can usually approximate $R'_{out}$ and $R_{in} \ll R_B$,

$$A'_i \approx A_i \quad \text{(Eqn. 11-3c)}$$

Output Resistance- $R_{out}$

$$R_{out} = \frac{1}{h_{ob} - \dfrac{h_{fb} \cdot h_{rb}}{h_{ib}}} \quad \text{(Eqn. 11-4a)}$$

Because $R_{out} \gg R_c$, $R'_{out}$ is simply $R_c$.

$$R'_{out} \approx R_c \quad \text{(Eqn. 11-4b)}$$

CB Example

To determine component values for our amplifier's quiescent point, we followed the same procedure as for the CE amplifier. Assuming $R_B = 5$ K$\Omega$, $R_E = 560$ $\Omega$, $V_{cc} = 20$ v, $V_C = 10$v, and $I_c = 1.5$ ma, we calculated these bias resistor values.

$$\frac{R_2}{R_1 + R_2} \cdot 20 - 1.5 \cdot 10^{-3} \cdot 560 = 0.7$$

$$\frac{R_2}{R_1 + R_2} = \frac{1.54}{20} = 0.077 \quad \text{(Eqn. 11-5a)}$$

$$\frac{R_1 \cdot R_2}{R_1 + R_2} = 5000 \quad \text{(Eqn. 11-5b)}$$

After we divided Eqn. 11-5b by Eqn. 11-5a, we found that,

$$R_1 = 61 \text{ K}\Omega \approx 56 \text{ K}\Omega \text{ as the nearest 5\% value.}$$

After we inserted $R_1 = 56$ K$\Omega$ in Eqn. 11-1a, we also found that,

$R_2 = 5.1 \, K\Omega \approx 5.6 \, K\Omega$ as the nearest 5% value.

Lastly, we calculated $R_c$,

$$R_C = \frac{10}{1.5 \cdot 10^{-3}} = 6.67 \, K\Omega = 6.8 \, K\Omega \text{ as the nearest 5\% value.}$$

The completed circuit was,

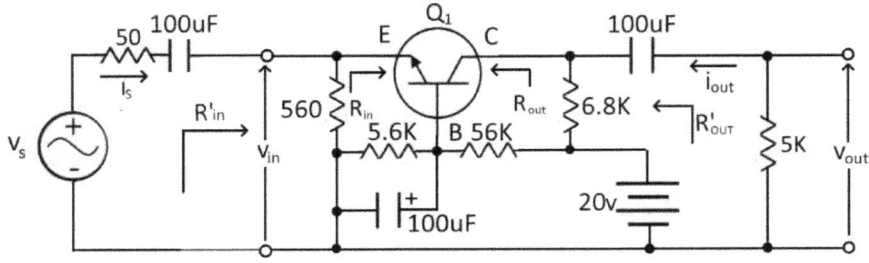

We used the CE h-parameter charts for the 2N3904 and Eqns. 8-4 to calculate the CB h-parameters.

For $I_c$ = 1.5 ma,

$h_{ie} = 2600 \, \Omega$
$h_{re} = 1.1 \cdot 10^{-4}$
$h_{fe} = 140$
$h_{oe} = 1.2 \cdot 10^{-5}$ μmhos

and

$h_{ib} = 18 \, \Omega$
$h_{rb} = 1.11 \cdot 10^{-4}$
$h_{fb} = -0.9929$
$h_{ob} = 8.5 \cdot 10^{-8}$ μmhos.

With these and the generalized hybrid Eqns.8-15, we calculated the amplifier performance characteristics.

$$A_v = \frac{v_{out}}{v_{in}} = \frac{-h_{fb} \cdot R_C}{(h_{ib}) \cdot (1 + h_{ob} \cdot R_C) - h_{fb} \cdot h_{rb} \cdot R_C}$$

$$A_v = \frac{-(-0.9929 \cdot 6800)}{(18) \cdot (1 + 8.5 \cdot 10^{-8} \cdot 6800) - (-0.9929 \cdot 1.11 \cdot 10^{-4} \cdot 6800)} = 360$$

$$A'_v = \frac{v_{out}}{v_s} = \frac{19}{19+50} \cdot 360 = 99$$

Moving to the input resistance,

$$R_{in} = h_{ib} - \frac{h_{rb} \cdot h_{fb}}{\left(h_{ob} + \frac{1}{R_C}\right)} = 18 - \frac{1.11 \cdot 10^{-4} \cdot -0.9929}{\left(8.5 \cdot 10^{-8} + \frac{1}{6800}\right)} = 19\ \Omega$$

Since $R_{in} \ll R_B$ (19 $\Omega$ $\ll$ 5000 $\Omega$),

$$R'_{in} = R_{in} = 19\ \Omega$$

Now we found the current gain to be,

$$A_i = \frac{i_c}{i_E} = \frac{h_{fb}}{(1 + h_{ob} \cdot R_C)} = \frac{-0.9929}{(1 + 8.5 \cdot 10^{-8} \cdot 6800)} = -0.9923$$

$$A'_i = \frac{i_{out}}{i_s} = \frac{5000}{19 + 5000} \cdot \frac{6.8}{6.8 + 5} \cdot (-0.9923) = -0.5718$$

and output resistance,

$$R_{out} = \frac{1}{h_{ob} - \frac{h_{fb} \cdot h_{rb}}{h_{ib}}} = \frac{1}{8.5 \cdot 10^{-8} - \frac{-0.9926 \cdot 1.11 \cdot 10^{-4}}{18}} = 165K\Omega$$

$$R_{out} = h_{ib} = 165K\Omega$$

$$R'_{out} = R_{out} \parallel R_c = 6.8K\Omega$$

As we noted above, the CB amplifier has low input resistance and high output resistance. While the current gain is less than 1, the voltage gain is high, easily on par with the CE amplifier.

Here are the test results.

| | Expected | Simulation | Breadboard |
|---|---|---|---|
| $A_v$ | 360 | 305 | 330 |
| $A_i$ | 0.9929 | 0.9933 | ~1 |
| $R'_{in}$ | 18 $\Omega$ | 22 $\Omega$ | 15 $\Omega$ |
| $R'_{out}$ | 6.8K $\Omega$ | 6.8 K$\Omega$ | 6.65 K$\Omega$ (used 5% 100 K$\Omega$ resistors) |

Given the range of h-parameters and 5% component tolerances, the values were well within the expected values. Regarding $R'_{out}$, it was equal to the value of $R_C$, indicating that $R_{out}$ was exceedingly large (calculated 165 KΩ). Similarly, the low value of $R_{in}$ meant that the value of $R_E$ had a negligible effect on the value of $R'_{in}$.

In the next chapter, we consider the low-frequency response of BJT amplifiers.

# Chapter 12 – BJT Amplifier Low-Frequency Response

Recall that when studying the CE amplifiers, we used RC coupling and bypassed the emitter resistor with a capacitor. We assumed that the coupling and bypass capacitors were short circuits at the signal frequency of 1000 Hz. In the basic CE amplifier, for example, we used 47 µF capacitors with a calculated *capacitive reactance* (signal resistance) of

$$R_c = \frac{1}{2\pi f C} = \frac{1}{2\pi \cdot 1000 \cdot 47 \cdot 10^{-6}} = 3.4 \,\Omega$$

This value is much less than their associated resistors, so our assumption that it was a short circuit was a good one. We refer to frequencies where circuit capacitors are short circuits as *mid-frequencies*, $f_{mid}$. We refer to the associated gains as *mid-frequency gains*, $A_{vMid}$ or $A_{iMid}$. In the previous amplifier examples, we would refer to 1000 Hz as a mid-frequency and the gains as mid-frequency gains. Were we to divide the gain by the mid-frequency gain, the result would be the normalized gain, $A_{vN}$ or $A_{iN}$. Normalized gain is 1 at mid-frequencies and either greater than 1 or less than 1 elsewhere,

Audio amplifiers typically have reduced gain above and below mid-band. In this chapter, we will focus on the reduced gain below mid-band. We refer to the measure of this reduced gain as the amplifier's *low-frequency response*. In the next chapter, we will consider reduced gain above mid-band, the amplifier's *high-frequency response*.

In the CE amplifier of Chapter 9, consider what happens if we lower the signal frequency from a $f_{mid}$ of 1000 Hz to 5 Hz. The reactance of the 0.47 µF capacitors increases to 677 $\Omega$ ($\frac{1}{2\cdot\pi\cdot f\cdot C}$), large enough to disrupt our original voltage gain calculations. Increased coupling capacitor reactance will inhibit signal passage between stages lowering voltage gain. Increased bypass capacitor reactance will increase the emitter resistance, introducing negative feedback and lowering gain further. The combined effect is lower stage gain as operating frequencies descend below $f_{mid}$. We will now investigate the BJT amplifier's low-frequency response, first due to the coupling capacitor and then the bypass capacitor.

## Coupling Capacitor Effects

To study coupling capacitor effects on low-frequency response, we will use the single-stage CE amplifier shown below.

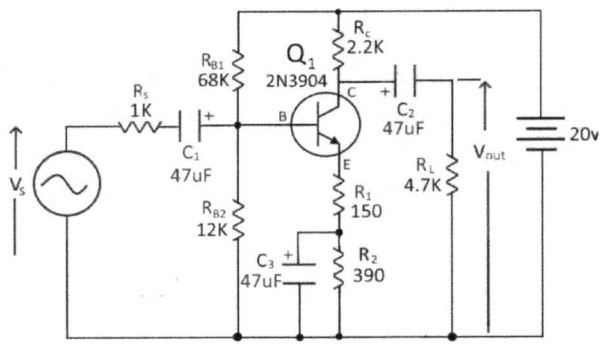

To get an idea of the extent of $C_2$'s effect on low-frequency response, we replace the BJT with a modified hybrid equivalent.

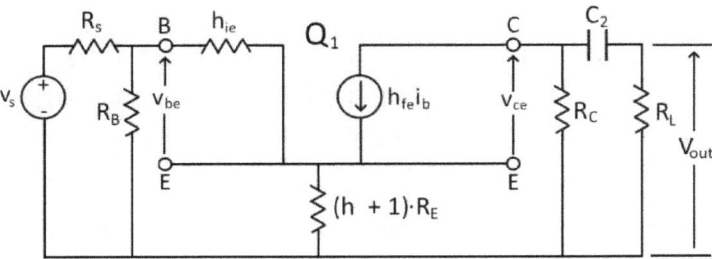

We assume that the CE hybrid parameters are,

$h_{ie}$ = 1200 Ω
$h_{re}$ = 0
$h_{fe}$ = 150
$h_{oe}$ = 0 mhos

We also assume that $C_1$ and $C_3$ are large enough that we can ignore their low-frequency effects. To determine how much the coupling capacitor $C_2$ affects the low-frequency response, we isolate it along with the resistances on either side.

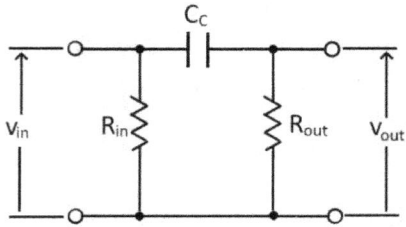

We will use the terms *input* and *output* to refer to the $v_{in}$ side and $v_{out}$ side of the capacitor, respectively. $R_{in}$ and $R_{out}$ are the equivalent resistances on either side of $C_c$. In our example, only one resistor is on each side, so $R_{in} = R_C$ and $R_{out} = R_L$.

Having determined the input and output resistances, we apply Thevenin's Theorem giving,

$$R_{low} = R_{in} + R_{out} \quad \text{(Eqn. 12-1)}$$

Thus, the circuit becomes a voltage divider and a *high-pass filter*.

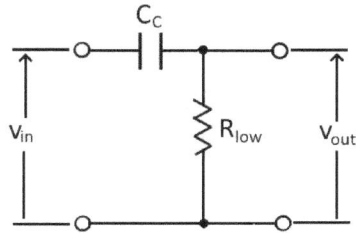

High-frequency signals pass unimpeded, while low-frequency signals attenuate as the capacitor's reactance increases. The impedance of the coupling circuit is,

$$Z_{cc} = R_{low} + \frac{1}{2\pi f C_2} i$$

Note: Because $Z_{cc}$ is an impedance, we write it as a complex number with a real ($R_{low}$) and an imaginary ($\frac{1}{2\pi f C_2}$) term. The imaginary symbol "i" is the value $\sqrt{-1}$. The main thing to know is that the magnitude of the complex number a + bi is Mag(a + bi) = $\sqrt{a^2 + b^2}$.

Using the voltage divider law, the circuit's gain is,

$$A_v = \frac{V_{out}}{V_{in}} = \frac{R_{low}}{R_{low} + \frac{1}{2\pi f C_C} i}$$

The voltage gain is the magnitude of the right side of the equation,

$$Mag(A_v) = \frac{R_{low}}{\sqrt{R_{low}^2 + (\frac{1}{2\pi f C_C})^2}} = \frac{1}{\sqrt{1 + (\frac{1}{2\pi f R_{low} C_C})^2}}$$

We define the *cutoff frequency* $f_{low}$ as the frequency when $A_v$ is down by 70.7%; that is when the denominator is $\sqrt{2}$. This occurs when,

$$f_{low} = \frac{1}{2\pi R_{low} C_C} \quad \text{(Eqn. 12-2)}$$

We can rewrite the magnitude above as,

$$Mag(A_v) = \frac{1}{\sqrt{1 + (\frac{f_{low}}{f})^2}}$$

We now compute $f_{low}$ for coupling capacitor $C_2 = 47\ \mu F$ of our example circuit,

$$R_{low} = R_c + R_L = 2200 + 4700 = 6900\ \Omega$$

$$C_C = C_2$$

and,

$$f_{low} = \frac{1}{2\pi R_{low} C_2} = \frac{1}{2\pi \cdot 6900 \cdot 47 \cdot 10^{-6}} = 0.49\ Hz$$

To summarize, the mid-frequency gain of the CE amplifier is,

$$A_{vMid} = \frac{R_C}{R_E} = \frac{2200||4700}{150} \approx \frac{1500}{150} = 10$$

The amplifier gain attenuated by the coupling capacitor $C_2$ is the product of $A_{vMid}$ and that introduced by the capacitor or,

$$A_v = A_{vMid} \cdot \frac{1}{\sqrt{1 + \left(\frac{f_{low}}{f}\right)^2}}$$

To study the frequency response, we calculate the normalized gain $A_{vN}$ by dividing by $A_{vMid}$,

$$A_{vN} = \frac{1}{\sqrt{1 + \left(\frac{f_{low}}{f}\right)^2}}$$

Here, we have plotted this relationship with standard axes.

With the axes as they are, we have entirely lost gain detail below 10 Hz!

Note: $A_{vMid} = 1$ and $\log(1) = 0$, so in the plot, $A_{vMid}$ gain is 0 DB.

To avoid this, we made two changes. First, we modify the gain equation as shown below.

$$A_{vN} = 20 \cdot \log\left(\frac{1}{\sqrt{1 + \left(\frac{f_{low}}{f}\right)^2}}\right) = -20 \cdot \log\left(\sqrt{1 + \left(\frac{f_{low}}{f}\right)^2}\right)$$

$$= -10 \cdot \log\left(1 + \left(\frac{f_{low}}{f}\right)^2\right)$$

$$A_{vN} = -20 \cdot \log\left(\frac{f_{low}}{f}\right) \quad (Eqn.\, 12 - 3)$$

Second, we change to a logarithmic frequency scale on the frequency axis. After applying these changes, the $A_{vN}$ plot is much more informative.

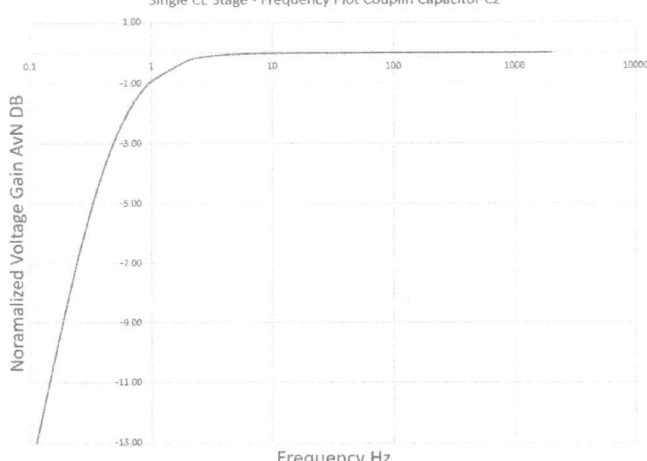

Single CE Stage - Frequency Plot Couplin Capacitor C2

We can now clearly see the attenuation detail below 10 Hz!

Recall that, at the cutoff frequency of 0.49 Hz, the voltage gain attenuation was 70.7% of the mid-band gain. As a fraction, this is 0.707. Thus, the DB value of 0.707 is,

$$A_{vN} = 20 \cdot Log(0.707) = 20 \cdot -0.15 = -3DB$$

Here is a revised plot showing the -3 DB attenuation at 0.49 Hz.

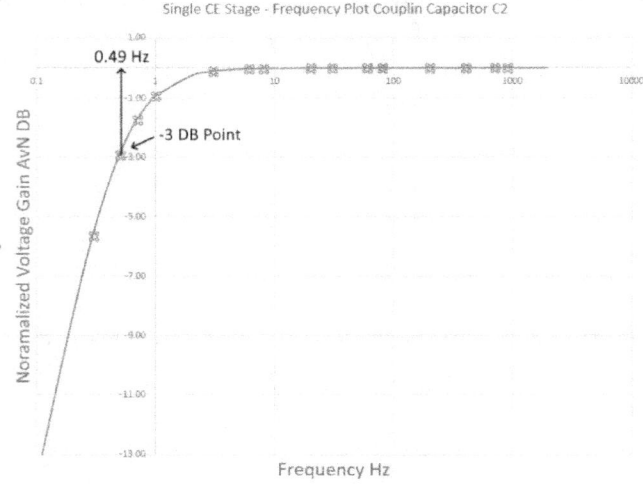

Single CE Stage - Frequency Plot Couplin Capacitor C2

We use the phrase *-3 DB point* to indicate the frequency when voltage gain has attenuated to 70.7% of its mid-band value.

Typically, BJT audio amplifiers have a -3 DB point above and below $f_{mid}$. We define the amplifier's *bandwidth* as the difference between the high and low -3 DB points. Another fact we find helpful is the *attenuation per frequency decade* (change of frequency by a factor of 10). This is the calculation for a factor of 10 attenuation.

$$A_{v10} = 20 \cdot Log(0.1) = 20 \cdot -1 = -20 \; DB$$

The ultimate slope of a single coupling capacitor plot is 20 DBs/decade. We can use this fact to sketch the $C_2$ frequency response. To see how, first assume $C_2$ = 1µF, which makes $f_{low}$ = 23 Hz. Starting at $f_{low}$ 23 Hz, draw a line with a 20 DB per decade slope as shown below.

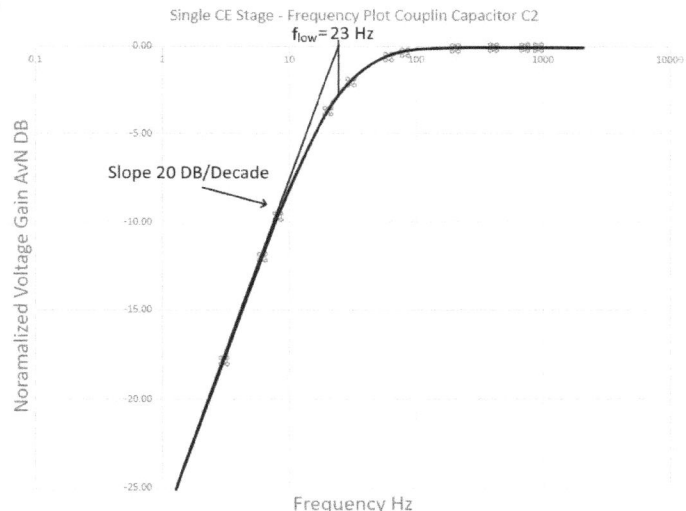

We can then sketch the $C_2$ low-frequency response starting at 0 DB, passing through the -3 DB point, and asymptotically approaching the 20 DB per decade line.

When designing an amplifier, we choose the coupling capacitor size that provides an acceptable $f_{low}$ cutoff frequency. If the circuit contains multiple coupling capacitors, the overall effect is the product of their attenuations or the sum of their DB attenuations. The combined effect increases the overall cutoff frequency and reduces the width of the mid-band from the low-frequency end. The plot below shows the effect of two stages identical to the single stage above.

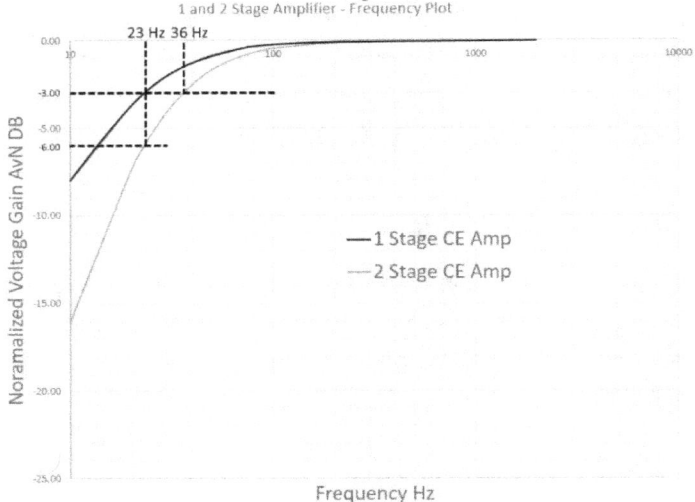

By adding a second coupling capacitor, the 23 Hz cutoff point is now down 6 DBs, and the -3 DB point has moved upward to 36 Hz. Adding a stage has noticeably lowered the low-frequency response. Increasing the coupling capacitors to 100 µF moves the cutoff frequency downward, readily compensating!

Our original CE amplifier included two coupling capacitors (C1 and C2). We will find it instructive to calculate the attenuation due to the other coupling capacitor, C1, as the base and collector resistances are different. For reference, here again is the circuit.

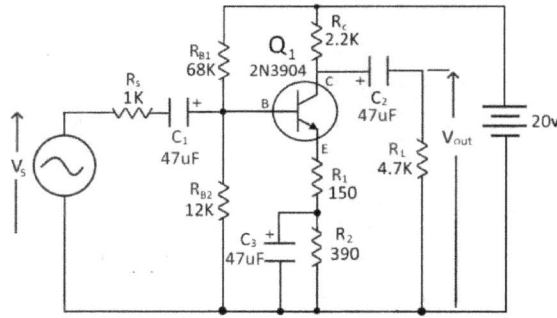

On the input side, the equivalent resistance $R_{in}$ is simply $R_s = 1000\ \Omega$. On the output side, $R_{out}$ is the parallel combination of bias resistance $R_B$ and the input resistance of $Q_1$. $R_B$ is the parallel combination of $R_{B1}$ and $R_{B2}$, which is 10,200 $\Omega$. Assuming that $C_3$ is a short circuit at the $C_2$'s cutoff frequency, then,

$$R_{in} = h_{ie} + (h_{fe} + 1) \cdot R_1 = 1200 + 151 \cdot 150 = 23{,}850 \ \Omega$$

So,

$$R_{out} = \frac{10{,}200 \cdot 23{,}850}{10{,}200 + 23{,}850} = 7144 \ \Omega$$

$R_{in} = 1000 \ \Omega$ and then,

$$R_{low} = R_{in} + R_{out} = 1000 + 7144 = 8144 \ \Omega$$

Calculating the cutoff frequency, we get,

$$f_{low} = \frac{1}{2 \cdot \pi \cdot R_{low} \cdot C_1} = \frac{1}{2 \cdot \pi \cdot 8144 \cdot 47 \cdot 10^{-6}} = 0.42 \ Hz$$

Combining this result with that above for $C_2$, we have,

$$A_{vN} = -20 \cdot \left[ \log\left(\frac{0.46}{f}\right) + \log\left(\frac{0.42}{f}\right) \right]$$

When plotted, we have,

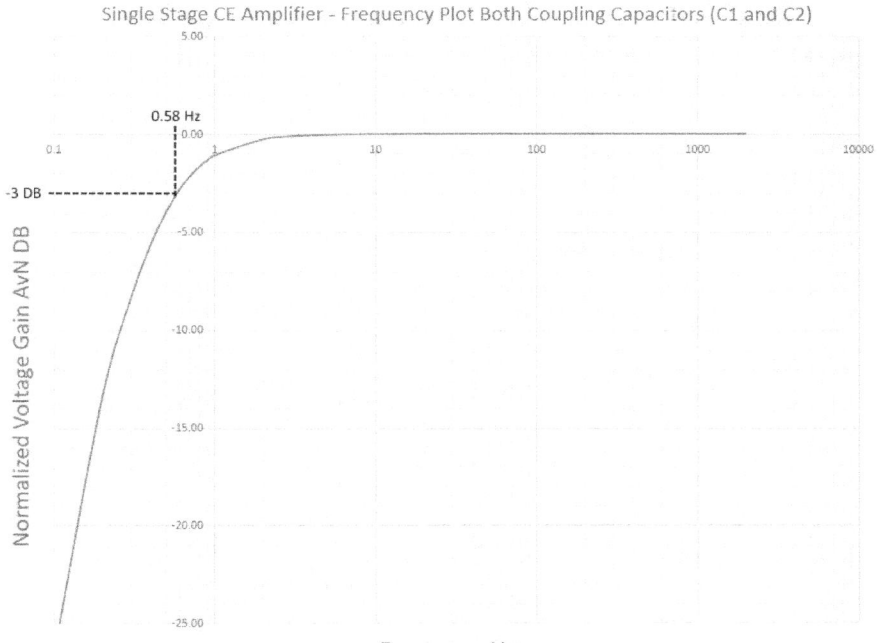

The effect of both coupling capacitors has moved the -3 DB point up to 0.58 Hz.

## Emitter Bypass Capacitor Effects

Next, we examine the effect of the bypass capacitor on low-frequency response. In the CE amplifier circuit of Chapter 9, the emitter resistance was partially bypassed; that is, we bypassed resistor R2 and not $R_1$. See below.

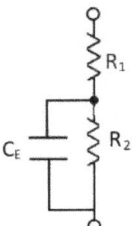

For our amplifier, the mid-band voltage gain was,

$$A_{vMid} = -\frac{R_L||R_C}{R_1} = \frac{2200||4700}{150} \approx \frac{1500}{150} = 10$$

As we lower the signal frequency, $C_E$'s reactance and the parallel combinations of $C_E$ and $R_2$ will increase. In turn, the total emitter resistance will increase, and the amplifier gain will decrease until $C_E$'s reactance is so large with respect to R2 that the gain $A_v$ is,

$$A_v = \frac{1500}{R_1 + R_2} = \frac{1500}{540} = 2.8$$

The frequency plot looks like this,

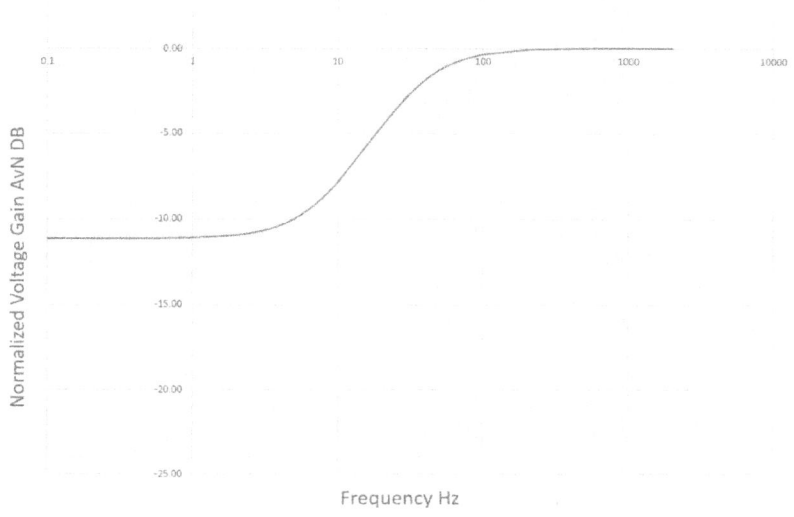

At about 100 Hz, the response begins a downward curve at 10 DB/decade, then at about 9 Hz, it begins to level out to $A_v$ = -11 DB (normalized gain of 0.28).

Our analysis begins by noting that voltage gain is,

$$A_v = \frac{R_C || R_L}{Z_E}$$

where $Z_E$ is the impedance of the emitter circuit below,

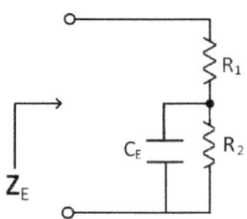

A simple analysis of this series-parallel circuit yields,

$$Z_E = R_1 + \frac{R_2 \cdot \dfrac{1i}{2\pi f C_3}}{R_2 + \dfrac{1i}{2\pi f C_3}} = R_1 + \frac{R_2 i}{2\pi f R_2 C_3 + i} = \frac{2\pi f R_1 R_2 C_3 + (R_1 + R_2)i}{2\pi f R_2 C_3 + i}$$

We are interested in the magnitude of $Z_E$, which is,

$$Mag(Z_e) = \frac{\sqrt{(2\pi f R_1 R_2 C_3)^2 + (R_1 + R_2)^2}}{\sqrt{(2\pi f R_2 C_3)^2 + 1}} \quad \text{(Eqn. 12-4)}$$

We can show that the -3 DB point occurs when $Mag(Z_E) = \lambda = 1.414 \cdot R_1$. If we set Eqn. 12-4 equal to $\lambda$ and solve for $C_3$, we find that,

$$C_3 = \frac{1}{2 \cdot \pi \cdot f_L \cdot R_2} \cdot \sqrt{\frac{\lambda^2 - (R_1 + R_2)^2}{R_1{}^2 - \lambda^2}} \quad \text{(Eqn. 12-5)}$$

where $f_L$ = the desired -3 DB frequency.

Consider now our amplifier.

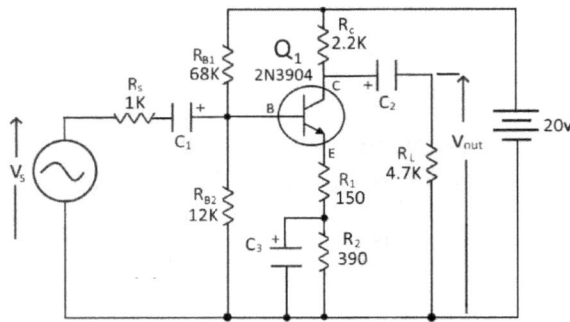

Suppose we want the lower -3 DB frequency to be 10 Hz. The design procedure below provides a way to find values for all the capacitors.

1. We ignore the coupling capacitors for the moment and use Eqn. 12-5 to calculate $C_3$.

$$\lambda = 1.414 \cdot R_1 = 212$$

$$C_3 = \frac{1}{2 \cdot \pi \cdot f_L \cdot R_2} \cdot \sqrt{\frac{\lambda^2 - (R_1 + R_2)^2}{R_1{}^2 - \lambda^2}}$$

$$= \frac{1}{2 \cdot \pi \cdot 10 \cdot 390} \cdot \sqrt{\frac{212^2 - (150 + 390)^2}{150^2 - 212^2}} = 140 \ \mu F$$

The nearest standard value for $C_3$ is 150 μF.

2. We next choose the cutoff frequencies for coupling capacitors $C_1$ and $C_2$ at 1/10 of $C_3$'s or 1 Hz. Now, use the $f_{low}$ formulas to calculate $C_1$ and $C_2$.

$$R_{low} = R_s + (h_{ie} + (h_{fe} + 1)$$
$$\cdot R_1) || R_B = 1000 + (1200 + ((150 + 1) \cdot 150)) || 10200$$
$$= 8144\Omega$$

$$C_1 = \frac{1}{2 \cdot \pi \cdot f_{low} \cdot R_{low}} = \frac{1}{2 \cdot \pi \cdot 1 \cdot 8144} = 19.5 \approx 20 \, \mu F$$

$$R_{low} = R_c + R_L = 2200 + 4700 = 6900\Omega$$

$$C_2 = \frac{1}{2 \cdot \pi \cdot f_{low} \cdot R_{low}} = \frac{1}{2 \cdot \pi \cdot 1 \cdot 6900} = 23 \, \mu F \approx 20 \, \mu F$$

3. Plot the results.

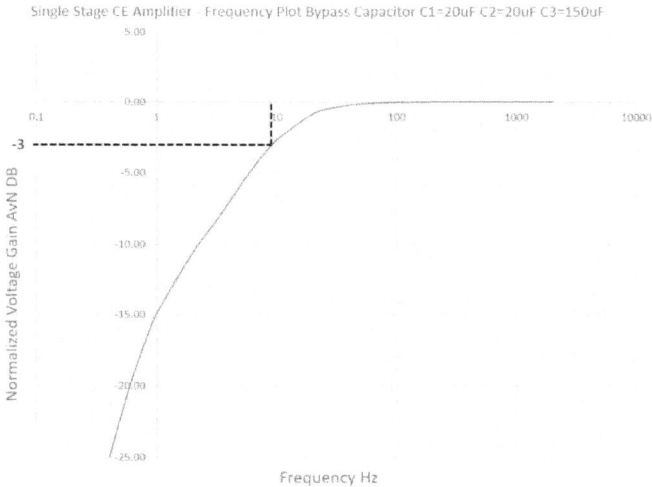

The low-frequency response looks good with the -3 DB point near 10 Hz and the response falling off smoothly after that.

For coupling capacitors in CC and CB amplifiers, we use the same approach as with the CE. We find the equivalent resistance input and output of the capacitor, add them together, and use the desired cutoff frequency $f_{low}$ to calculate its value. For the emitter bypass capacitor, we choose a value that is a short circuit at the calculated $f_{low}$.

Now that we know how to choose capacitors to set the low-frequency response, in the next chapter, we examine the high-frequency response of BJT amplifiers.

# Chapter 13 – BJT Amplifier High-Frequency Effects

In this chapter, we examine factors that limit BJT amplifier high-frequency response. Of these, two are primary contributors.

(1) BJT's internal capacitances create low-pass filters that limit high-frequency response.

(2) The value of $h_{fe}$ falls off above the BJT's transition frecuency (call it $f_T$). When choosing a BJT, we choose one with a $f_T$ far above the amolifier's operating frequency. The $f_T$ of the 2N3904 is 300 MHz, so that we can ignore it for our audio frequency analysis. With power BJTs, $f_T$ is typically less than 30 MHz, and we cannot do it.

Regarding BJT internal capacitances, the figure below shcws them as capacitors added to the BJT hybrid model. (We have assumed that both $h_{re}$ and $h_{oe}$ are zero.)

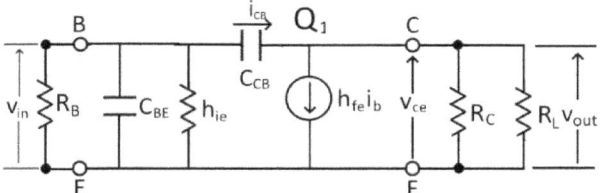

Shown are two capacitors representing (1) the capacitance between the collector to the base ($C_{CB}$ above) and (2) the capacitance between the base and emitter ($C_{BE}$ above). For the 2n3904 BJT, $C_{CB}$ is ~4 pF, and $C_{BE}$ is ~8 pF.

$C_{CB}$ introduces negative feedback from the collector output to the base input. The effect of $C_{CB}$ is to reduce voltage gain as the signal frequency increases and $C_{CB}$'s reactance decreases. $C_{BE}$ placed across the input shunts the input signal $v_{in}$ to the common at high frequencies.

The effect of $C_{BE}$ is much less than that of $C_{CB}$. The reason is that $C_{CB}$ affects both the input <u>and</u> output of a BJT amplifier, while $C_{BE}$ affects only the input. To see how we first examine $C_{CB}$'s effect on the input.

$$i_{CB} = 2\pi f C_{CB} \cdot (v_{in} - v_{out}) \quad (Eqn.\ 13 - 1)$$

where $\dfrac{1}{2\pi f C_{CB}}$ is the capacitive reactance of capacitor $C_{CB}$ at frequency f. Since $i_{CB}$ << $h_{fe} \cdot i_b$, then

$v_{out} = -A_v \cdot v_{in}$ and,

$$2\pi f C_{CB}\left(v_{in} - (-A_v \cdot v_{in})\right) = 2\pi f C_{CB} v_{in}(1 + A_v) = 2\pi f \cdot (1 + A_v)C_{CB} \cdot v_{in}$$

This suggests that the current $i_{CB}$ drained from the base is the same as that produced by a capacitor of the value $C_h = (A_v + 1) \cdot C_{CB}$ connected from the base to the emitter.

The input effect of $C_{CB}$ is an example of the *Miller Effect*, named in honor of John Milton Miller. In the 1920s, Miller discovered the effect of a similar capacitance in vacuum tube triodes. As we will see shortly, the Miller Effect is the major limiting factor in determining the high-frequency response of triodes and BJTs.

As noted above, $C_{CB}$ also affects the output. To see how, we start again with Eqn.13-1.

$$i_{CB} = 2\pi f C_{CB} \cdot (v_{in} - v_{out})$$

This time, we extract $v_{out}$ and find that,

$$i_{CB} = 2\pi f C_{CB} \cdot v_{out} \left(\frac{v_{in}}{v_{out}} - 1\right) = -2\pi f C_{CB} \cdot v_{out}\left(\frac{1}{A_v} + 1\right)$$
$$= -2\pi f \cdot \left(\frac{A_v + 1}{A_v}\right) C_{CB} \cdot v_{out}$$

This suggests that the same current would flow in a capacitor of value $\left(\frac{A_v+1}{A_v}\right) C_{CB}$ from collector to emitter.

Below is the equivalent circuit with $C_{CB}$ replaced by the Miller Effect capacitors.

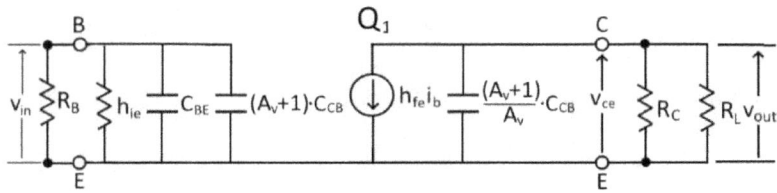

For $A_v > 10$, the primary high-frequency effect is due to the capacitance $(A_v+1) \cdot C_{CB}$ on the input. For $A_v < 10$, we must consider the effects of both, but as we will see, the effects are minor unless we are working with frequencies above 1 MHz.

To analyze the input effect, we write the Thevenin Equivalent of the input circuitry and find that,

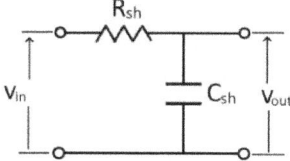

where $R_{sh} = R_B || h_{ie}$ and $C_{sh} = (1 + A_v) \cdot C_{CB} + C_{BE}$.

Unlike the high-pass filter of a capacitive coupling circuit, we have a *low-pass filter*. Low-frequency signals pass unimpeded, while capacitance $C_{sh}$ shunts high-frequency signals to the emitter. Applying the same analysis as before, we obtain,

$$A_v = \frac{v_{out}}{v_{in}} = \frac{1}{1 + 2\pi f R_{sh} C_{sh} i} = \frac{1}{1 + (\frac{f}{f_{high}})i}$$

where $f_{high} = \frac{1}{2\pi R_{sh} C_{sh}}$.

The magnitude of the mid-band gain of our amplifier is,

$$A_{vN} = \frac{1}{\sqrt{1 + \left(\frac{f}{f_{high}}\right)^2}} \text{ or } A_{vN} = -20 \cdot \log\left(\frac{f}{f_{high}}\right) \quad (Eqn.\ 13-2)$$

To illustrate the capacitive effects, we will apply them to the amplifier below.

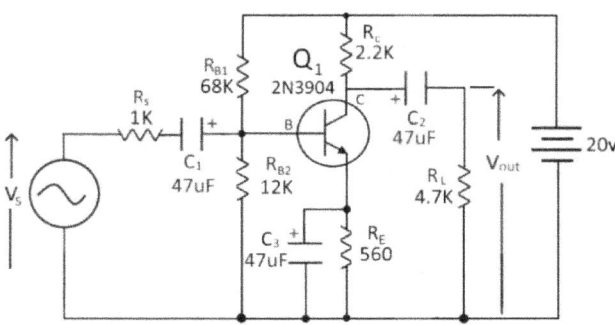

The first step is to calculate the shunt resistance.

$$R_{sh} = R_s || R_B || h_{ie} = 518 \; \Omega$$

The data sheet for the 2N3904 shows $C_{obo}$ = 4 pF, which is the maximum output capacitance between the collector and base with the emitter open-circuited. Since any capacitance between the base and emitter would only decrease this, we can assume the $C_{CB}$ < $C_{obo}$ and use $C_{CB}$ = 4 pF for our calculation. We can make a similar argument to $C_{BE}$ = 8 pF based on the 2N3904 specification of $C_{ibo}$ maximum = 8 pF. For the mid-band voltage gain $A_v$, we use $A_v = \dfrac{R_c || R_L \cdot h_{fe}}{h_{ie}} = \dfrac{1500 \cdot 150}{1200} = 187$. Then,

$$C_{sh} = 188 \cdot 4 + 8 = 760 \; pF$$

and the worst-case input cutoff frequency is,

$$f_{high-input} = \frac{1}{2\pi R_{sh} C_{sh}} = \frac{1}{2\pi \cdot 518 \cdot 761 \cdot 10^{-12}} = 404 \; KHz$$

Since the output value of $C_{CB}$ is small and its cutoff frequency high (>25 MHz), we ignore it. The high-frequency plot for $C_{CB}$ = 4 pf is,

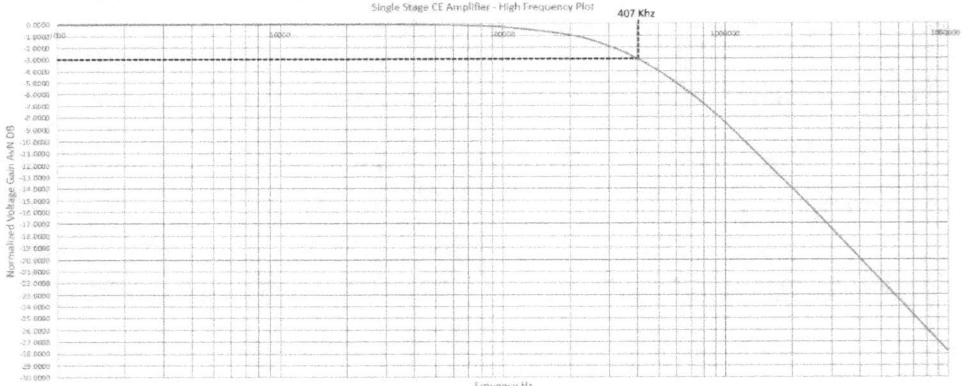

The simulated value of the cutoff frequency was ~320 KHz. The SPICE value of $C_{CB}$ was 3.2 pF. We measured the $C_{CB}$ value of the breadboarded amplifier as 0.88 pF, and the high-frequency cutoff was an impressive 3.3 MHz! If we were building an audio amplifier, the 2N3904 high-frequency response would be more than adequate. Below is the breadboard frequency response plot.

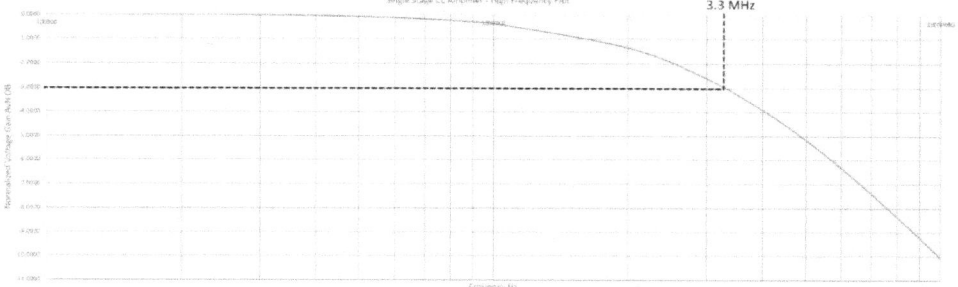

This is an excellent time to introduce the idea of the *gain-bandwidth product*. We start with the hybrid representation of a CE amplifier.

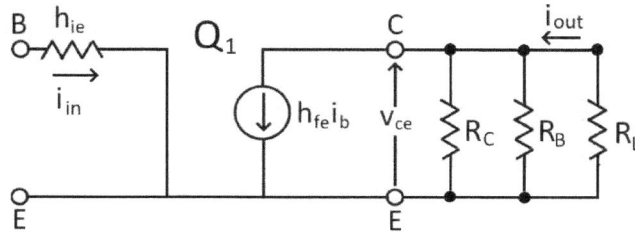

R_c is the collector resistor of the amplifier stage, R_B is the bias resistor of the following stage, and R_L is the input load of the following stage. From Eqn. 9-5b, we can deduce the mid-band current gain is,

$$A_{imid} = \frac{i_{out}}{i_{in}} = \frac{h_{fe} \cdot R_{sh}}{R_L} \quad \text{where } R_{sh} = R_C||R_B||R_L \quad (\text{Eqn.13-3})$$

To study the high-frequency effects, we add $C_{CE}$ (the collect-emitter capacitance of the amplifier stage) and $C_L$ (the load capacitance of the following stage) to the output.

We then define $C_{sh} = C_{CE} + C_L$.

The circuit looks like this,

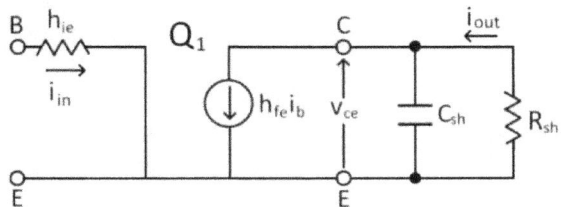

Were we to analyze the circuit's high-frequency response as we did earlier, we would find that the cutoff frequency is,

$$f_{high} = \frac{1}{2\pi R_{sh} C_{sh}} \quad \text{(Eqn. 13-4)}$$

From Eqn.13-3, we have $R_{sh} = \frac{A_{imid} \cdot R_L}{h_{fe}}$

Substituting this in Eqn. 13-4, we get,

$$f_{high} = \frac{h_{fe}}{2\pi A_{imid} R_L C_{sh}}$$

Multiplying by A$_{imid}$ gives us the product of voltage gain and bandwidth,

$$A_{imid} \cdot f_{high} = \frac{h_{fe}}{2\pi R_L C_{sh}}$$

Since $A_{vmid} = A_{imid} \cdot \frac{R_{sh}}{R_L}$, we can conclude that,

$$A_{vmid} \cdot f_{high} = \frac{h_{fe}}{2\pi C_{sh}} = f_T \quad \text{(Eqn. 13-5)}$$

We call the right-hand side the *transition frequency* and designate it f$_T$. We now see that the so-called *gain-bandwidth product* equals the transition frequency f$_T$.

As an example, f$_T$ for the 2N3904 is 300 MHz. To check the validity of Eqn.13-5, given A$_v$ =102,

$102 \cdot f_{high} = 300$ MHz and $f_{high} = \frac{300}{102} = 2.94$ MHz which agrees well with the results above, which were 3.3 MHz.

We find Eqn. 13-5 is applicable when selecting a BJT for a particular purpose because it estimates the maximum frequency of operation given an expected voltage gain. A germanium BJT typically has high internal capacitances and a low f$_T$, which limits its use in high-frequency RF applications. The availability of high f$_T$ silicon BJTs today provides a better gain in RF applications.

The complete frequency response (called a *Bode Plot*) of our breadboarded CE amplifier appears below.

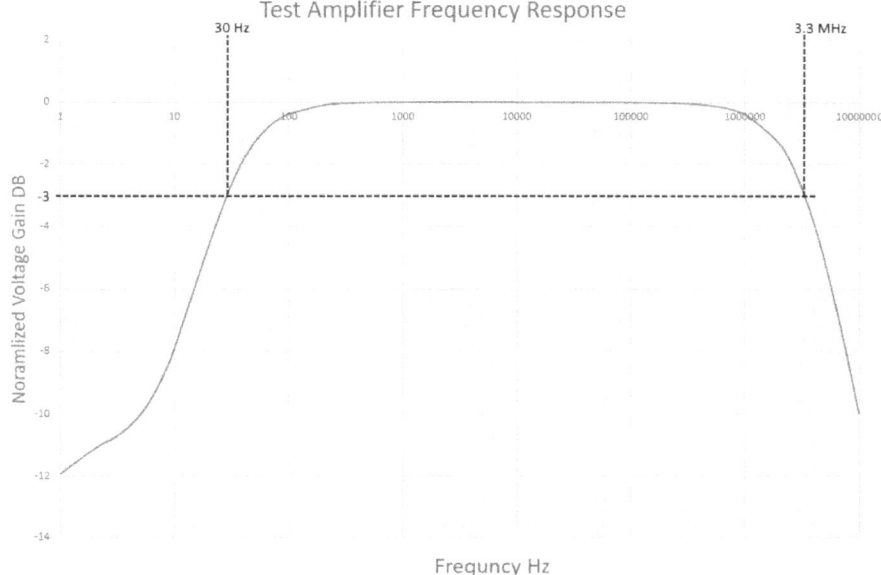

Based on the -3 DB points, its response is 30 Hz to 3.3 MHz. Because the low-frequency cutoff is so low, we equate bandwidth to the high-frequency cutoff. So, we say its bandwidth is ~3.3 MHz.

Before we leave high-frequency effects, we should consider the high-frequency response of CC and CB amplifiers. The critical difference between these configurations and the CE is that they do not suffer from the Miller Effect.

In the case of the CC amplifier, the collector's constant voltage means there is no signal to feedback via internal capacitance. Thus, if the design objective is current gain, the CC configuration offers significant high-frequency advantages.

The simulated CC amplifier below has a $f_{high}$ of 33 MHz compared to the CE version's 3.3 MHz!

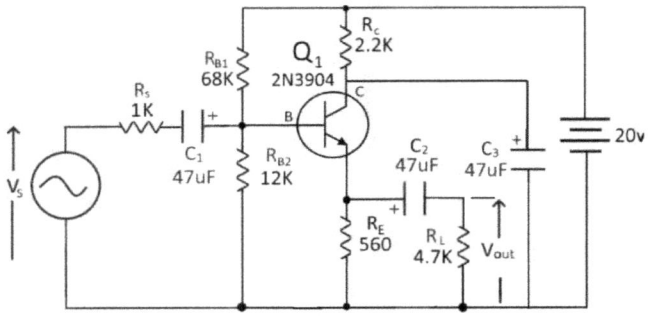

The CB amplifier avoids the Miller Effect because its base connected to common acts as a shield that prevents signal feedback via parasitic capacitance $C_{CB}$. For this reason, CB amplifiers are frequently used as *very high frequency* (VHF) amplifiers.

Consider the CB amplifier below.

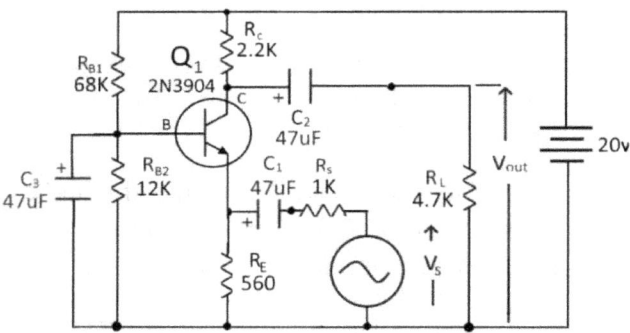

The CB's simulated cutoff frequency $f_{high}$ is 58 MHz compared to the CC's 33 MHz and the CE's 3.3 MHz. In addition, the CB's voltage gain at higher frequencies is larger than either the CE or CC.

In the net chapter, we apply frequency response theory to multistage amplifiers.

# Chapter 14 – Multistage Amplifiers

BJT amplifiers often consist of multiple stages. In this chapter, we will study a *two-stage, cascaded amplifier* configuration. In Chapter 8, we found that we could readily obtain voltage gains of more than 100 with a single CE BJT stage. We also found that $A_v$ highly depended on $h_{fe}$ that could range in production BJTs like the 2N3904 from 100 to 300. In non-critical applications, we might accept the resulting variation in $A_v$, but for more critical applications, we need a better approach. We could, for example, divide the gain between two gain-stabilized stages using unbypassed emitter resistors.

Cascaded stage designs introduce two complications.

(1) BJT amplifiers have a non-zero input resistance $R_{in}$. Thus, when cascading stages, we must consider the effect of the subsequent stage's input resistance on the gain of the previous stage.

(2) Cascaded stages reduce the BJT amplifier's bandwidth. Therefore, to maintain a specified bandwidth, we must increase the bandwidth of each stage. While we can easily do this at low frequencies, it becomes more difficult at high frequencies.

We will interconnect CE amplifier stages to demonstrate the design approach for a cascaded audio amplifier. To achieve an overall voltage gain $A_v$ of 100, we will cascade two partially unbypassed emitter CE amplifiers, splitting the gain equally between them. We will assume the output load $R_L$ is 4700 Ω and input source resistance $R_s$ is 1000 Ω. As this is to be an audio amplifier, we will specify a frequency response of ±2 DB from 20 to 20,000 Hz. Here is our two-stage amplifier.

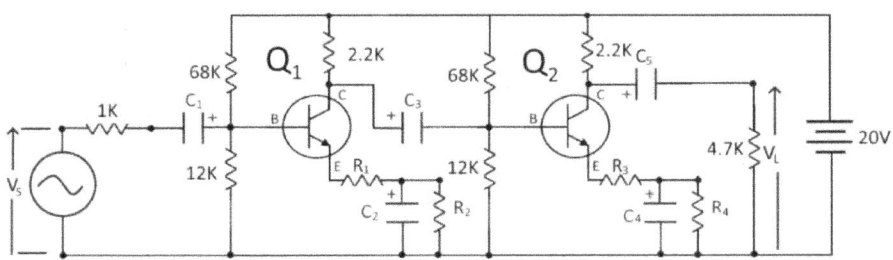

We will start with a higher target mid-band voltage gain $A_{vMid}$ of 121, recognizing that trimming it down later to 100 is easy to do; we increase the bypassed emitter resistance. Lastly, we will split the gain equally between stages, making $A_{vMid}$ = 11 for each.

To start, we will assume that the coupling and bypass capacitors are short circuits at audio frequencies. Later, we will address low-frequency effects and select capacitor values. Since we are designing for audio frequencies and we are using $f_T$ = 300 MHz BJTs, we need not be concerned about high-frequency response.

We begin at the $Q_2$ stage and work backward. We know that $Q_2$'s collector load resistance is,

$$R_{cQ2} = \frac{2200 * 4700}{(2200 + 4700)} = 1500\Omega$$

Therefore, $R_3$ is $R_{cQ2}$ / 11 = 1500 / 11 = 136 Ω. The nearest, smaller 5% value for $R_3$ = 120 Ω. $R_4$ is then 560 – 120 = 440 Ω. The nearest 5% value for $R_4$ = 470 Ω.

Note: By choosing the smaller standard value for $R_3$, we picked a value that increased voltage gain. Here again, picking the higher gain option is a good strategy. It is better to miss high than low!

Next, we calculate $R_{inQ2}$.

$R_{inQ2}$ = 68K||12K || ($h_{ie}$ + ($h_{fe}$ + 1) · $R_{cB}$ = 10.2K || (1200 + (150 + 1) · 120 = 6647 Ω. $Q_1$'s

and the collector load resistance $R_{cQ1}$ ,

$$R_{cQ1} = \frac{2200 * 6647}{(2200 + 6647)} = 1864\Omega$$

$R_1$ then is,

$$R_1 = \frac{R_{cQ1}}{11} = \frac{1864}{11} = 169 \ \Omega$$

We choose the nearest, smallest 5% value, $R_1$ = 150 Ω, making $R_2$ = 560 – 150 = 410 Ω. We choose the nearest 5% value of $R_2$ = 390 Ω.

In summary, these were the emitter resistors,

$Q_1$
$R_1$ = 150 Ω

$R_2 = 390\ \Omega$
Q2
$R_3 = 120\ \Omega$
$R_4 = 470\ \Omega$

Here is the circuit thus far with all resistors in place.

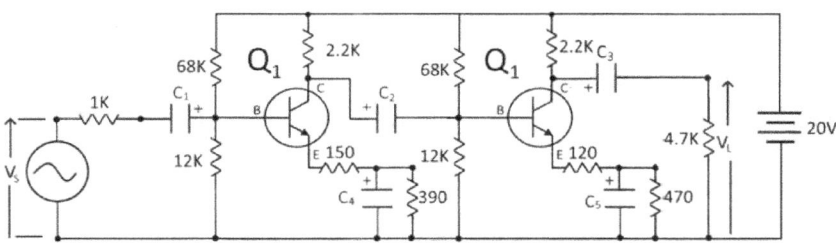

As a last step, we check the voltage drop at the voltage divider input to be sure it does not lower the voltage gain significantly:

$R_{inA} = 68K || 12K || (h_{ie} + (h_{fe} + 1) \cdot R_{cB} = 10.2K || (1200 + (150 + 1) \cdot 150 = 7117\ \Omega$.

The voltage divider factor is 7117 / (1000 + 7117) = 0.89. Starting with $A_v = 121$ and applying this factor, we calculate $121 \cdot 0.89 = 108$, wh ch is above our specification of 100.

Here are the test results.

|  | Expected | Simulation | Breadboard |
|---|---|---|---|
| K | 0.89 | 0.87 | 0.94 |
| $A_{vQ1}$ | 12.4 | 11.2 | 12.9 |
| $A_{vQ2}$ | 12.5 | 10.1 | 8.4 |
| $A_v$ | 138 | 98 | 102 |
| $R'_{out}$ | 6.8K $\Omega$ | 6.8 K$\Omega$ | 6.65 K$\Omega$ (used 5% 100 ‹$\Omega$ resistors) |

Though individual gains varied from design values, the overall gain was close to the specified overall gain $A_{vMid}$ of 100. In further tests, we substituted BJTs with various combinations of $h_{fe}$. We observed only minor char ges in overall voltage gain. The design achieved our goal of a stable, multistage CE amplifier with a mid-band voltage gain of 100.

One lesson to learn goes back to the strategy of opting for more, not less, voltage gain at each step. As is evident from the results above, gain values always seemed less than predicted. Had we not opted for higher gain at each step, we would have missed the target $A_v$ = 100. Then, we would have had to look for places to increase gain. In this example, this would have been easy enough to do by lowering $R_1$ or $R_3$. In other situations, especially when working at RF frequencies, finding ways to achieve higher gain might not be so easy!

Knowing the mid-band gain, we are ready to address the frequency response of ±2 DB from 20 to 20,000 Hz. From our previous experience with this circuit and the 2N3904 BJT, we know that high-frequency response is no problem. We will check it later to be sure. We must select the coupling and emitter bypass capacitors to achieve a low-frequency response of -2 DB or better at 20 Hz. We will use the procedure in Chapter 12 to estimate their values.

Our first step is to set a value for the cutoff frequency. We want the gain to be down no more than 2 DB at 20 Hz. If we used a cutoff frequency of 20 Hz, the gain would be down to at least 9 DB because we have two stages plus the source capacitor $C_1$. This is an example of *bandwidth shrinkage* that occurs with multistage amplifiers. To compensate for this (at least partially), we make the following adjustment:

$$f_{low} \ (new) = f_{low} \ (old) \cdot \sqrt{2^{\frac{1}{n}} - 1} \quad \text{(Eqn. 14-1)}$$

where n = number of coupling capacitors.

For our two-stage amplifier, our new cutoff frequency will be,

$$f_{low} = 20 \cdot \sqrt{2^{\frac{1}{3}} - 1} \approx 10 \text{ Hz}$$

Therefore, we should design for 3 DB down at 10 Hz. We are now ready to follow the procedure in Chapter 12.

1. Ignoring the coupling capacitors for the moment, we use Eqn. 12-5 to calculate $C_4$.

$$\lambda = 1.414 \cdot R_1 = 1.414 \cdot 150 = 212$$

$$C_4 = \frac{1}{2 \cdot \pi \cdot f_L \cdot R_2} \cdot \sqrt{\frac{\lambda^2 - (R_1 + R_2)^2}{R_1{}^2 - \lambda^2}}$$

$$= \frac{1}{2 \cdot \pi \cdot 10 \cdot 390} \cdot \sqrt{\frac{212^2 - (150 + 390)^2}{150^2 - 212^2}} = 135 \; \mu F$$

The nearest standard value on the high side is 150 µF. We will use this value for bypass capacitors $C_4$ and $C_5$.

Note: By choosing the high side, we were opting to push the -3 DB point lower. A smaller capacitor would have raised the -3 DB point. We choose the next higher bypass capacitor value in such situations.

2. We next choose the cutoff frequencies for coupling capacitors $C_1$, $C_2$, and $C_3$ at 1/10 of $C_4$ and $C_5$ or 1 Hz. We use the $f_{low}$ formulas of Chapter 12 and calculate $C_1$, $C_2$, and $C_3$.

$$R_{low1} = R_s + \left( h_{ie} + \left( h_{fe} + 1 \right) \right.$$
$$\left. \cdot R_1 \right) \middle|\middle| R_B = 1000 + \left( 1200 + ((150 + 1) \cdot 150) \right) \middle|\middle| 10200$$
$$= 8144 \; \Omega$$

$$C_1 = \frac{1}{2 \cdot \pi \cdot f_{low} \cdot R_{low1}} = \frac{1}{2 \cdot \pi \cdot 1 \cdot 8144} = 20$$
$$\approx 25 \; \mu F \text{ (nearest convenient value)}$$

$$R_{low2} = R_c + R_B || R_{inQ2} = 2200 + 10200 || (1200 + (151) \cdot 120) = 8875 \; \Omega$$

$$C_2 = \frac{1}{2 \cdot \pi \cdot f_{low} \cdot R_{low2}} = \frac{1}{2 \cdot \pi \cdot 0.8 \cdot 8875} = 18$$
$$\approx 25 \; \mu F \text{ (nearest convenient value)}$$

$$R_{low3} = R_c + R_L = 2200 + 4700 = 6900 \; \Omega$$

$$C_3 = \frac{1}{2 \cdot \pi \cdot f_{low} \cdot R_{low3}} = \frac{1}{2 \cdot \pi \cdot 0.8 \cdot 6900} = 23$$
$$\approx 25 \; \mu F \text{ (nearest convenient value)}$$

The completed circuit is,

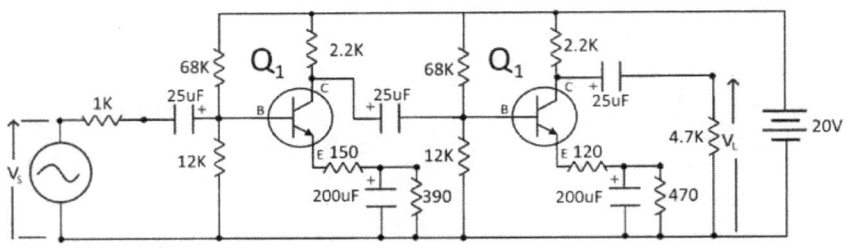

### 3. Plot the simulated results.

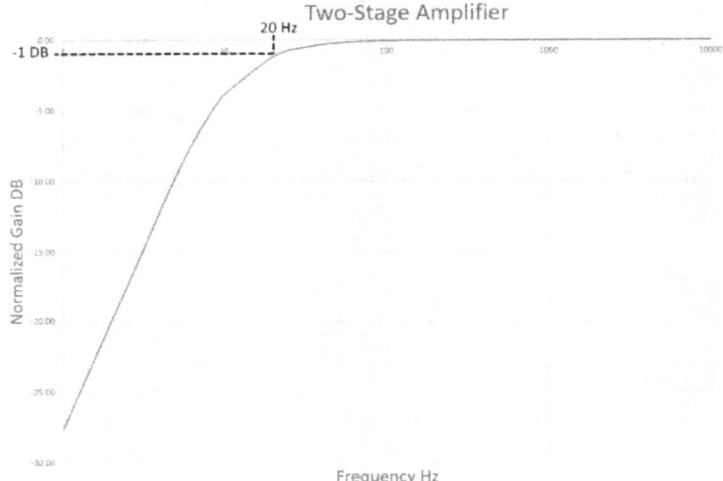

At 20 Hz, voltage gain is down 1 DB, well within the specified value of down 2 DB! The breadboard version was down 1.2 DB at 20 Hz but was still within specifications.

As suggested earlier, the high-frequency response was unlikely to be a problem, but we should check it to be sure. The Miller Effect capacitors on Q1 and Q2 limit the high-frequency response. We analyze these points below using the maximum $C_{CB}$ for the 2N3904 (4 pF).

For $Q_1$:

$$A_v = \frac{R_C||R_B||(h_{ie} + (h_{fe} + 1) \cdot R_{1Q2})}{R_{1Q1}} = 11$$

$$\left(\frac{1}{R_{high}}\right)^{-1} = \left(\frac{1}{1000} + \frac{1}{10200} + \frac{1}{1200+(150+1)\cdot150}\right)^{-1} = 877 \ \Omega$$

$$C_{high} = (A_v + 1) \cdot C_{CB} + C_{BE} = 12 \cdot 4 \cdot 10^{-12} + 2 \cdot 10^{-12} = 50\, pF$$

$$f_{high} = \frac{1}{2\pi \cdot 877 \cdot 50 \cdot 10^{-12}} = 3.63\ \text{MHz}$$

For $Q_1$:

$$A_v = \frac{R_C \| R_L}{R_{1Q2}} = 12 \left(\frac{1}{R_{high}}\right)^{-1} = \left(\frac{1}{1200} + \frac{1}{10200} + \frac{1}{1200+(150+1)\cdot120}\right)^{-1} = 1017\ \Omega$$

$$C_{high} = (A_v + 1) \cdot C_{CB} + C_{BE} = 13 \cdot 4 \cdot 10^{-12} + 2 \cdot 10^{-12} = 28\, pF$$

$$f_{high} = \frac{1}{2\pi \cdot 1017 \cdot 54 \cdot 10^{-12}} = 2.90\ \text{MHz}$$

Even the combined high-frequency effects of both stages would not affect the high-frequency response at 20,000 Hz.

The circuit met the specifications ±2 DB 20-20,000 Hz set forth at the start.

Note: By choosing two unbypassed emitter stages to achieve a mid-band gain of 100, the design assured us that variations in BJT parameters and component values would not seriously affect the amplifier's performance.

With all this focus on voltage gain, we should not lose sight of the fact that BJTs are current-driven devices. If we were to take advantage of this, we could directly couple one BJT to another and eliminate a coupling capacitor.

In Chapter 10, we developed a CC amplifier with an input load resistance to match a magnetic phono cartridge. While its input load resistance of 47 KΩ was correct, its gain of ~1 was unsuitable when considering a magnetic cartridge's meager output of 3-6 millivolts.

To address this, we will design a directly coupled two-stage amplifier. When we couple a CC amplifier directly to a CE amplifier, the former provides the correct load resistance for the magnetic phono cartridge, and the latter provides the needed voltage gain. Here is the hybrid model of our proposed circuit.

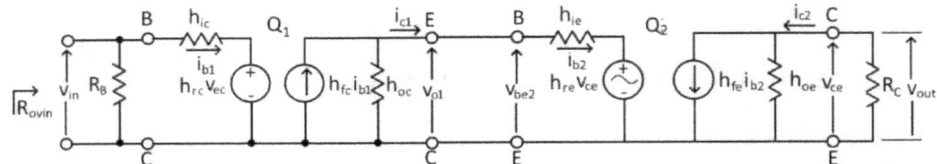

The analysis is easy if we assume $h_{oe} = 0$, $h_{re} = 0$, and the $v_{ec1} = v_{bc2} = v_{in}$. The modified hybrid circuit is then,

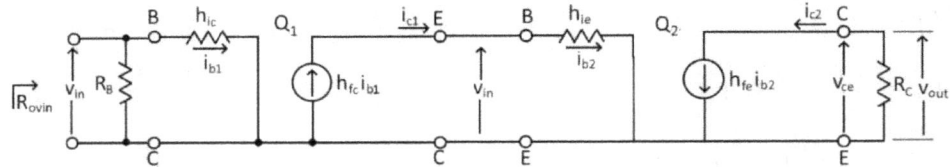

Because the CC stage's current output directly feeds the CE stage's base input base, it takes full advantage of $Q_1$'s current gain. What is more, we need no coupling capacitor!

Looking at $Q_2$, we see that,

$$i_{b2} = \frac{v_{in}}{h_{ie}} \text{ and } v_{out} = -h_{fe} \cdot i_{b2} \cdot R_C$$

Combining and simplifying, we find that,

$$A_v = \frac{v_{out}}{v_{in}} = -\frac{h_{fe} \cdot R_C}{h_{ie}}$$

We can now find the circuit's component values.

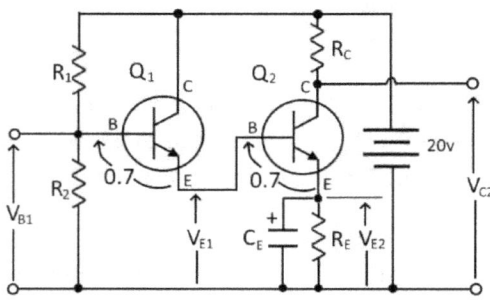

We assume the base-emitter voltages are the nominal 0.7 volts. We want the input resistance to be 47 KΩ, so we make $R_B$ = 47 KΩ. Further, we make $V_{C2}$ = 10

volts and allow 2 volts for $V_{E2}$. Also, we choose $I_{C2} = 4$ ma. With this information, we are ready to compute $R_1$ and $R_2$.

$$V_{B1} = 2 + 0.7 + 0.7 = 3.4\ v$$

$$R_B = \frac{R_1 \cdot R_2}{R_1 + R_2} = 47000 \quad and \quad V_{B1} = \frac{R_2}{R_1 + R_2} \cdot 20$$

Dividing the last two equations,

$$R_1 = \frac{47000}{3.4/20} = 276\ K\Omega \text{ use } 270\ K\Omega \text{ as nearest 5\% value.}$$

$$R_B = \frac{270K \cdot R_2}{270K + R_2} = 47000$$

Solving for $R_2$,

$$R_2 = 56,900\ K\Omega \text{ use } 56\ K\Omega \text{ as nearest 5\% value.}$$

Next, we calculate $R_E$ and $R_C$,

$$R_E = \frac{2}{4 \cdot 10^{-3}} = 500\ \Omega \text{ use } 470\ \Omega \text{ as nearest 5\% value.}$$

$$R_C = \frac{8}{4 \cdot 10^{-3}} = 2000\ \Omega \text{ use } 2200\ \Omega \text{ as nearest 5\% value.}$$

This is the completed circuit.

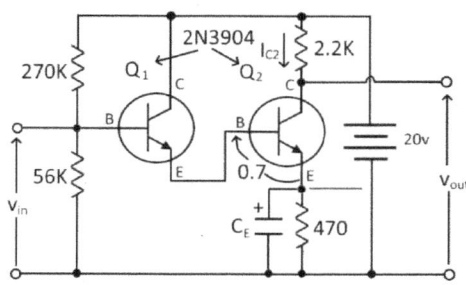

The voltage gain is,

$$A_v = \frac{v_{B1}}{v_{C2}} = -\frac{h_{fe} \cdot R_C}{h_{ie}} = -\frac{140 \cdot 2.2k}{2K} = 154$$

A test of the circuit yielded these results:

| | Expected | Simulation | Breadboard |
|---|---|---|---|
| $V_{C1}$ | 2 v | 2.2 v | 2.2 v |
| $V_{C2}$ | 10 v | 9.8 v | 9.75 v |
| $A_v$ | 154 | 121 | 150 |
| $R'_{in}$ | 47 KΩ | 47 KΩ | 51 KΩ (used 5% 100 KΩ resistors) |

All the values were reasonable given the wide range of h-parameters and the 5% components.

We can make a valuable variation of this circuit by moving the collector of $Q_1$ from $V_{cc}$ to the collector of $Q_2$. In theory, the configuration has a single base, emitter, and collector and appears to be a single BJT with an $h_{fe}$ equal to the product of the individual BJTs. See the figure below.

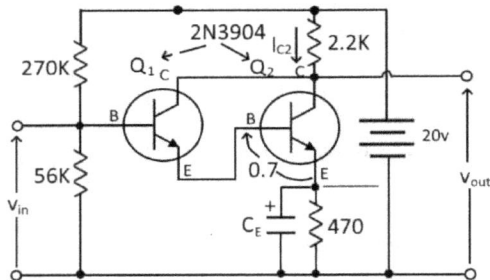

This interconnection of two BJTs is known as a *Darlington pair*, named for inventor and electrical engineer Sidney Darlington. While the directly coupled current gain is high, the voltage gain of the circuit is not because the value of the combined $h_{ie}$ is larger as well,

$$h_{ie} = h_{ieQ1} + (h_{fe} + 1) \cdot h_{ieQ1}$$

For the 2N3904, the Darlington pair $h_{ie}$ is,

$$h_{ie} = 2000 + (151) \cdot 2000 = 304 \ K\Omega$$

The value of voltage gain $A_v$ is,

$$A_v = -\frac{150^2 \cdot 2200}{304000} = 163$$

After changing the collector to the breadboard circuit, we measured $A_v = 160$, well within the predicted range.

Darlington pairs come in integrated packages and are available in small signal and high-current configurations. The BC-517 is a small signal Darlington pair. The TIP-120 (NPN) and TIP-125 (PNP) are complementary Darlington pairs for high-current motor control and audio power amplifiers.

With a voltage gain of 163, the amplifier would produce an output of ~0.5 volts with the worst-case phono cartridge output of 3 mv. By reducing the Darlington stage gain and adding another stage, we could make a more gain-stable version. See the circuit below.

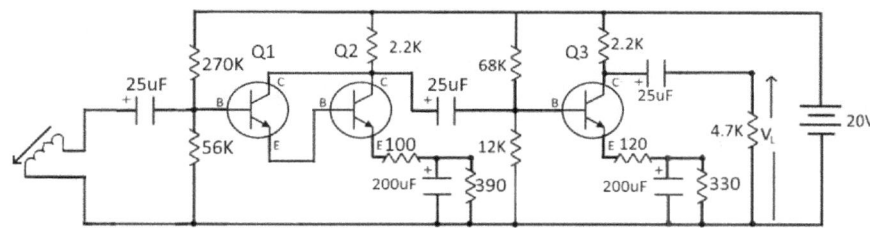

We have removed the bypass on a portion of $R_E$ and added a stage with an unbypassed emitter resistor. Using a mid-band gain of 13 for each stage, the overall gain is $13^2 = 169$, close enough to the original gain of 163.

We split the emitter resistors based on these calculations.

$$R_C || R_B || R_{inQ3} = 2200 || 10200 || 6676 = 1424 \ \Omega$$

$$R_{1Q2} = \frac{1424}{13} = 109 \approx 100 \ \Omega$$

$$R_{2Q2} = 470 - 100 = 370 \approx 390 \ \Omega$$

$$R_C || R_L = 2200 || 4700 = 1500 \ \Omega$$

$$R_{1Q3} = \frac{1500}{13} = 115 \approx 120 \ \Omega$$

$$R_{2Q3} = 470 - 120 = 350 \approx 330 \ \Omega$$

Using the identical capacitors as in the 2-stage amplifier, we had these breadboard results:

$V_{cQ1} = 10.3\ vdc$

$V_{cQ2} = 10.1\ vdc$

$A_v = 167$

Frequency Response

-1.4 DB at 20 Hz

-2.0 DB at 1.4 MHz

While these are excellent results, we have one more design consideration. Magnetic phono cartridges require a frequency compensation known as the RIAA equalization curve.

When making vinyl recordings, the record industry reduces low-frequency signals and increases high-frequency signals (called *preemphasis*). We must add a filter that modifies the amplifier's frequency response to compensate. We must decrease voltage gain by 20 DB per decade from 50 to 500 Hz. We then flatten the response until 2100 Hz when we decrease it again, 20 DB per decade. Of course, no practical filter can produce the exact response, so we must accept limitations.

The circuit below shows the circuitry we added to approximate the RIAA filter within the dotted rectangle.

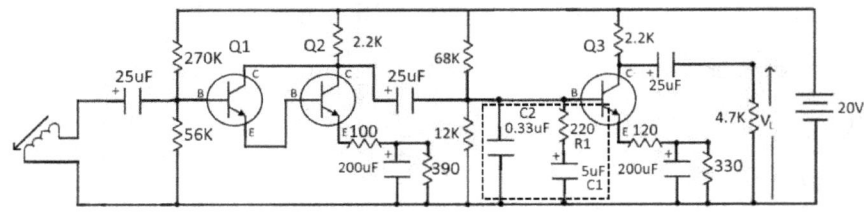

The combination of $R_1$ and $C_1$ provides the 50 to 500 Hz roll-off, while capacitor $C_2$ introduces the remaining 2100 Hz roll-off. When $C_1$ is a short circuit at mid-band frequency, the output resistance of $Q_2$ (~ 2.2 KΩ) and resistor $R_1$ function as a voltage divider, reducing $A_{vMid}$ from 100 to 10 (-20 DB). A simple calculation shows that we need $R_1$ = ~250 Ω. We then use the fact that,

$$C_1 \approx \frac{1}{2\pi \cdot f_{low1} \cdot R_1} = \frac{1}{2\pi \cdot 50 \cdot 250} = 12.7\ \mu F$$

$$C_2 \approx \cfrac{1}{2\pi \cdot \cfrac{R_{outQ1}}{\cfrac{R_{outQ1}}{R_1} + 1} \cdot f_{low2}} = \cfrac{1}{2\pi \cdot \cfrac{2200}{\cfrac{2200}{150} + 1} \cdot 2100} = 0.57 \ \mu F$$

At the breadboard stage, we found it necessary to finetune the response by trial and error with resistors and capacitors close to the calculated values. The final Bode plot of the breadboard version looked like this.

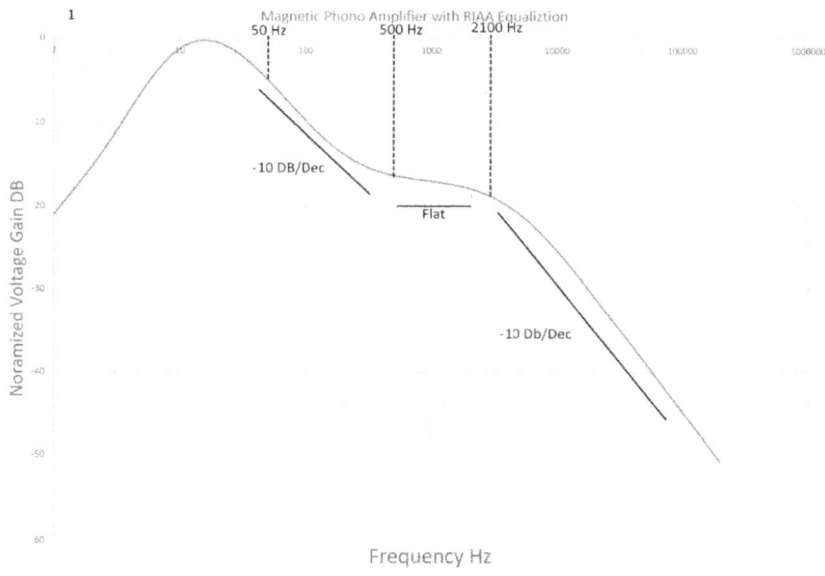

It is not perfect, but it is close enough for our purposes.

Note: Adding the filters lowered the mid-band gain, so we reduced the size of the unbypassed emitter resistors to compensate. With the change, we brought mid-band gain back up to ~160.

In the next chapter, we examine large signal or power amplifiers that can drive a speaker.

# Chapter 15 – BJT Single-Ended Power Amplifiers

One important capability of BJT amplifiers is they can amplify power. Power gain results from the fact that a BJT amplifier can amplify both voltage and current at the same time. To calculate power gain, we take the absolute value of the product of voltage and current gain; that is,

$A_P = |A_v \cdot A_i|$ where the symbol $|...|$ represents *absolute value*.

For example, our basic CE amplifier had a voltage gain $A_v$ = -95 and current gain $A_i$ = 45, making the power gain $A_P = |-93 \cdot 42| = 3906$. This means that the amplifier multiplies the input power by 3906 before delivering it to the load resistance. While this may sound impressive, it is not really, considering the few milliwatts output. For power amplifiers, we think in terms of watts, not milliwatts. With BJT power amplifiers, the power gain may be less, but the power output is considerably more!

In analyzing power amplifiers, the first thing that falls by the wayside is using small signal analysis. The voltage and current excursions are much larger than 10% of the quiescent point that we limited ourselves to with small signals analysis. We push BJT power amplifiers to 100% voltage-current excursions, which calls for *load line analysis*.

To explore maximum power output, we will first consider a CE power amplifier using a BJT with these ideal collector characteristics.

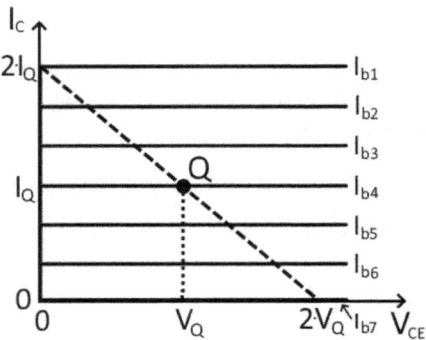

Note that (1) The collector current curves extend to the $I_C$ axis. (2) The collector current curve associated with minimum base current $I_{b7}$ falls on the $V_{CE}$ axis. (3) The spacing between collector current curves is regular.

The dotted line is the load line of maximum output. For this ideal case, a sinusoidal signal can traverse the entire load line from $2 \cdot I_Q$ to $2 \cdot V_{CE}$ with a maximum power output of

$$P_{OMax} = \frac{\Delta V \cdot \Delta I}{8} = \frac{(2 \cdot V_Q - 0) \cdot (2 \cdot I_Q - 0)}{8} = \frac{V_Q \cdot I_Q}{2} \text{ watts}$$

With no internal external resistances in the collector or emitter circuit, the total power consumed is $P_c = V_Q \cdot I_Q$. Therefore, the maximum efficiency is

$$\epsilon = \frac{\frac{V_Q \cdot I_Q}{2}}{V_Q \cdot I_Q} = 50\%$$

So, in an ideal world, the best we can hope for with a sing e-ended Class A amplifier is 50% efficiency. We will try to achieve this as we work through the CE amplifier designs below.

In an actual BJT amplifier, current passing from the emitter to the collector generates heat and wastes power. We call the equivalent resistance that produces this wasted power the BJT's saturation resistance $R_S$. If present, the collector resistor $R_c$ and the emitter resistor $R_E$ also waste power through heating. $R_S$, $R_C$, and $R_E$ impose an excursion limit alongside the collector current axis, as shown below.

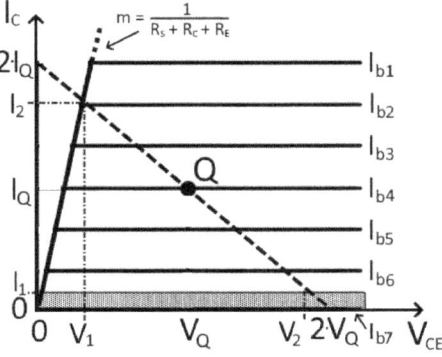

The limit consists of a line that passes through the origin (0, 0) with a slope $m = \frac{1}{R_S + R_c + R_E}$. The greater these resistances, the more the line leans to the right, and the greater the excursion limit imposed on the collector current.

On the collector voltage side of Q, the load line encounters a severe non-linearity as $I_c$ approaches the cutoff current, the shaded area in the figure above. Load line excursion into this region produces a severe flattening of the sinusoid signal and keeps $V_{CE}$ from reaching $2 \cdot V_q$.

The effect of both excursion limits is to decrease $\Delta V$ and $\Delta I$, their product, and power output. That is,

$$P_o = \frac{(V_2 - V_1) \cdot (I_2 - I_1)}{8} < \frac{V_Q \cdot I_Q}{2} = P_{oMax}$$

It follows that an actual BJT CE amplifier cannot achieve 50% efficiency.

We designate the power lost in the saturation resistance $R_s$ as $P_c$ and measure it in watts. We calculate it using the formula $P_c = I_Q \cdot V_Q$. In a small signal BJT amplifier, this power loss is trivial and of no concern. In power BJTs, the heat loss can be significant and cause an uncontrollable temperature rise known as *thermal runaway*. In this situation, heat rise causes more collector current, which causes more heat rise and more current, finally destroying the BJT.

To explore power loss and inefficiency, we will consider the small signal example below. (While power loss does not matter in this case, the presence of all components and the availability of collector characteristics for the 2N3904 make it an excellent example to study. The results apply to power BJTs.)

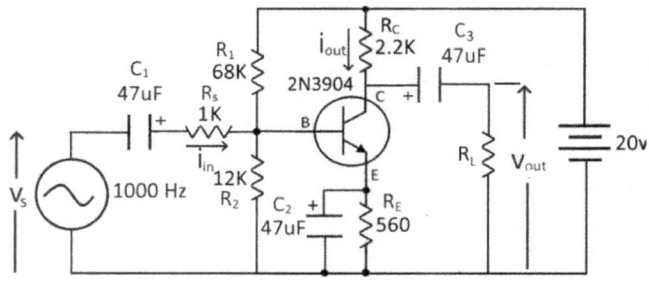

Given this circuit, our objective is to transfer maximum power to load resistor $R_L$.

Recall from Chapter 6 that the nominal collector current $I_c$ is 3.5 ma and $V_{CE}$ is 10 v. This makes $P_c = 0.0035 \cdot 10 = 35$ mw. The power loss in resistors $R_c$ and $R_E$ is then $(2200 + 560) \cdot 0.0035^2 = 34$mw. The total power loss in the collector circuit is 35 mw + 34 mw = ~69 mw.

The immediate question is whether an optimum output resistance value $R_{out}$ exists that will produce the maximum power transfer to the load resistance $R_L$. To investigate this, we simulated the circuit and used a parameter sweep on $R_L$ to measure power output. See the results below.

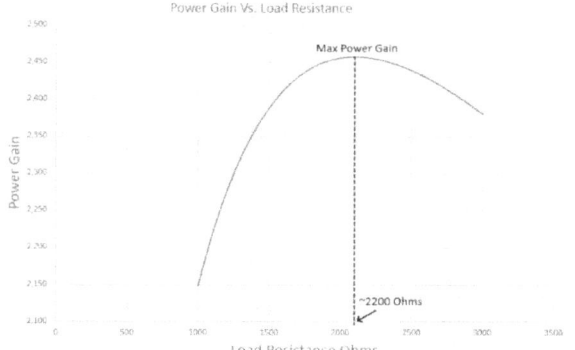

Note that maximum power output occurs when $R_L$ is $2200\Omega$. Ignoring $h_{oe}$, $R_{out}$ is also $2200\ \Omega$. This suggests that maximum power transfer occurs with $R_{out} = R_L$. This is not a surprising result as, in principle, maximum power transfer occurs when the load resistance of a device equals its output resistance. See Appendix C for the proof.

Digging a little deeper, the power delivered to the 2.2 K$\Omega$ load is ~6 mw, while the power lost in $R_c$, $R_E$, and $P_c$ is ~69 mw. Power transfer might be optimum, but 8.7% efficiency is poor!

Looking closely at the power losses, $R_c$ and $R_E$ contributec the most. Even with $R_c = R_L$, only 50% of the available power output is transferred to the load $R_L$. If we are to reach 50% efficiency, then one approach we could try is to minimize $R_E$ and make $R_c$ the load.

To explore this possibility, we will use a load line analysis of a CE amplifier and a TIP41 power BJT. Below are TIP41 collector characteristics.

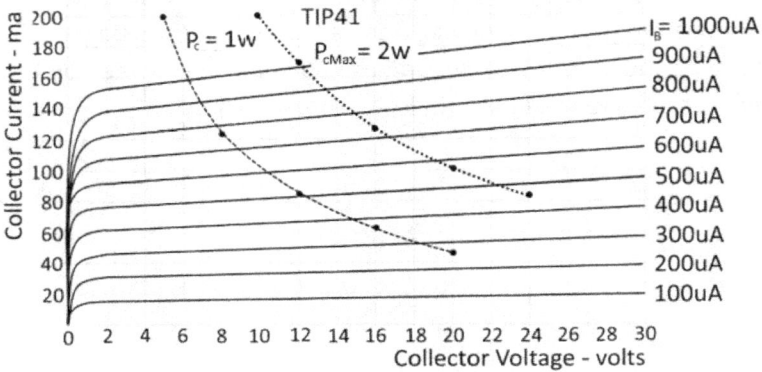

The plot shows TIP41 collector current vs. collector voltage for a range of base currents. The TIP41 specification gives a maximum of 2 watts of allowable power dissipation $P_{cMax}$ without a heat sink.

Since we will not be using a heatsink, we will limit $P_c$ to 1 watt to stay well within the 2-watt maximum. By plotting the $P_c = I_c \cdot V_{CE} = 1$-watt curve on the collector characteristics, we can choose a quiescent point on or below it to stay within the heat dissipation limit. Here is the basic CE circuit.

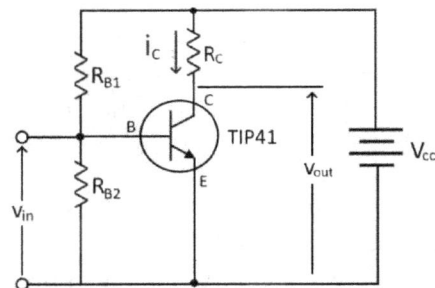

We eliminated the emitter resistor to improve efficiency and made the collector resistor $R_c$ also the load. We then drew the load line starting at the knee of the highest collector current, passing tangent to the $P_{cMax}$ curve, and striking the collector voltage axis at 24 volts. See the figure below.

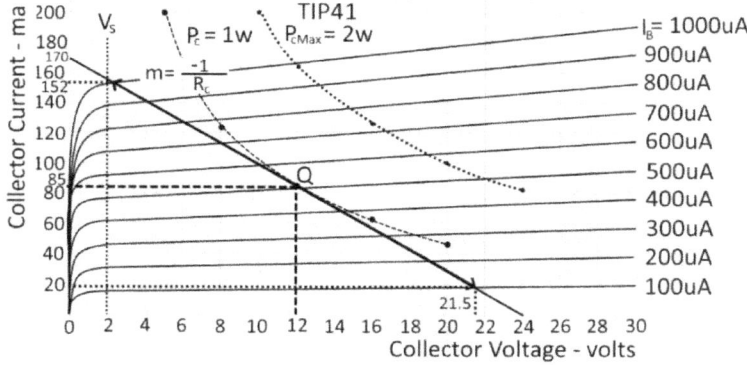

By including the tangent point of the $P_c$ = 1 w curve, we assure that collector dissipation will never exceed 1 watt. In addition, the excursion of the signal up and down the load line will always stay under 1 w dissipation.

Does choosing the load line this way provide maximum power output given the 1 w dissipation limit? Though not mathematical, we answer the question this way:

Power output is proportional to the product of the voltage and current excursion about Q. The load line's slope is $-1/R_c$. Changing the value of $R_c$ pivots the load line about Q. Increasing $R_c$ from its present value pivots the load line so that voltage excursion increases but the current excursion decreases. This is not likely to make their product (the power output) any greater. Decreasing $R_c$ pivots the load line so that the current excursion increases while the voltage excursion decreases, again unlikely to increase power output. In both cases, the operating point moves into the axis region, which further limits excursion and power output. We can assume that the maximum power output occurs at or near the load line described above.

We can calculate the value of $R_c$ from the load line slope this way.

Slope $m = \dfrac{-0.170}{24} = -7.083 \cdot 10^{-3}$ and $R_c = -\dfrac{1}{m} = 141\ \Omega$

The quiescent point at Q is (85ma, 12 volts) and $V_{cc}$ is the collector axis intercept of 24 volts.

The maximum power delivered to $R_c$ is,

$$P_{outmax} = \frac{\Delta i \cdot \Delta v}{8} = \frac{(.15-.018)\cdot(21.5-2)}{8} = 351\ mw$$

The power dissipated in $R_c$ is $141 \cdot 0.085^2 = 1.02$ w. $P_c$ is 1 watt, so the total power consumed is 2.02 watts. Therefore, the amplifier's efficiency is $\varepsilon = \frac{351}{2020} \cdot 100\% = 17\%$

This is an improvement, but still far short of the Class A maximum of 50%.

Improving efficiency is not the only issue with this circuit. If our goal is to drive a speaker with a resistance of 8 Ω, we are far from the 141 Ω load needed to maximize power output. An even more significant concern is that we cannot have DC current flowing through our speaker. Speakers are strictly AC devices!

We can solve the speaker load issue by substituting an audio power transformer for the collector resistor.

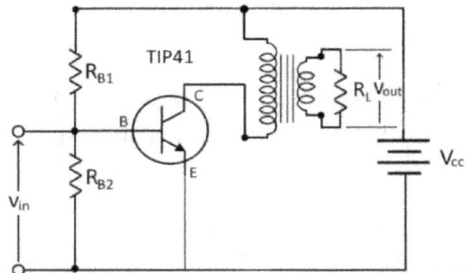

Using the transformer, we can match the low-resistance speaker (say 8 Ω) to the higher load resistance (141 Ω) of the TIP31. Using transformer coupling also solves the DC current issue by isolating the collector voltage from the speaker. Still, another benefit is that the DC resistance of the transformer primary is small, meaning much less power loss than with load resistor $R_c$.

For sinusoidal signals, the AC load $R'_L$ reflected through the transformer is determined by the equation,

$R'_L = (\frac{n_p}{n_s})^2 \cdot R_L = N^2 \cdot R_L$  (Eqn. 15-1) where $n_p$ is the number of turns in the transformer's primary winding and $n_s$ is the same for the secondary. N is the turns ratio $n_p/n_s$.

The design approach is as follows. Suppose now that our target dissipation is $P_c$ is 1 watt (still less than the 2 watts, the TIP31 maximum without heat sink) and the load is a speaker with $R_L = 8$ Ω. Here is a design procedure.

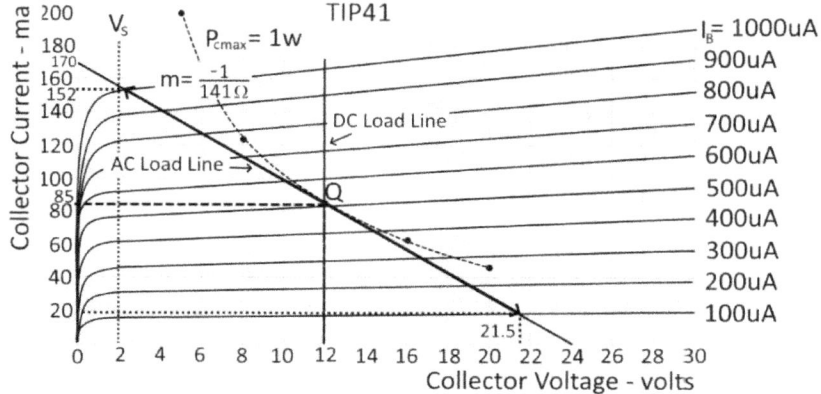

1. Draw on the collector characteristics the hyperbola representing $P_{cMax} = I_Q \cdot V_Q = 1$ watts. See above.

2. Draw the AC load line tangent to the $P_{cmax}$ curve that passes through the knee of the largest collector curve. See above.

3. Identify the quiescent point Q at the tangent point. $I_Q = 85$ ma.

4. Since there is no resistance in the collector circuit (assuming an ideal transformer), the DC load line is vertical and passes through the tangent point. This means that $V_{cc} = 12$ volts.

4. Identify the intercepts of the AC load line on the collector voltage and collector current axes. See above.

5. The voltage intercept to current intercept ratio is $R_{AC} = 24/.170 = 141\ \Omega$.

6. Find the turns transformer ratio $N = (141/8)^{0.5} = 4.2$.

Knowing Q, we can find the value of $R_{B1}$ and $R_{B2}$. The maximum bias current is 1 ma, which makes,

$$R_{B1} + R_{B2} = \frac{12}{10 \cdot 0.001} = 1200\ \Omega$$

Assuming $R_B = 1\ K\Omega$, the closest 5% bias resistors are,

$$R_{B2} = \frac{0.7 \cdot 1200}{12} = 70\ \Omega\ \text{(used 68 }\Omega\text{)}$$

$$R_{B1} = 1200 - 70 = 1130\ \Omega\ \text{(used 1200 }\Omega\text{)}$$

The completed circuit, then, is,

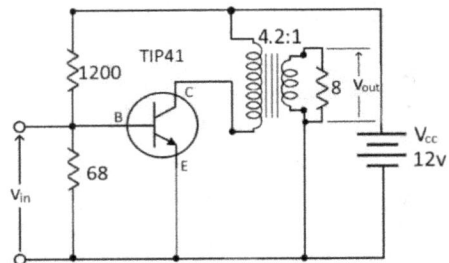

The power output is the same as before, 351 mw, but the total power loss is $P_c$ only or 1 w. Efficiency is now $\varepsilon = \frac{351}{1000} \cdot 100\% = 35\%$, much better!

Manufacturers rarely provide collector characteristics for their BJTs, so we must use an alternative procedure. The procedure below gives reasonable results, provided the resistance of the transformer primary is small.

1. Calculate a $P_{cmax} = 2 \cdot P_{out}$. Be sure that $P_{cmax}$ is less than the maximum dissipation allowed for the BJT, given the heat sink provided.

2. Choose a value for $V_{cc}$ such that $2 \cdot V_{cc} \le V_{CEmax}$, the maximum collector-emitter voltage allowed for the BJT.

3. Calculate $V_{max} = 2 \cdot V_{cc} - V_s$ where $V_s$ is the collector saturation voltage, usually about 2 volts.

4. Set $I_Q = \frac{2 \cdot P_{cmax}}{V_{max}}$.

5. Set $R_{AC} = \frac{V_{max}^2}{4 \cdot P_{cmax}}$.

To check this procedure, we can use it for the example above, assuming no collector characteristics were available.

1. Assume our goal is a power output $P_{out} = 350$ mw. $2 \cdot 350 = 700$ mw $< 1$ w $= P_{cmax}$.

2. Choose $V_{cc} = 12$ v less than $V_{CEmax} = 40$ v.

3. $V_{max} = 2 \cdot 12 - 2 = 22$ v.

4. Set $I_Q = \frac{2 \cdot 1}{22} = 90$ ma (compared to 83 ma).

5. Set $R_{AC} = \frac{22^2}{4 \cdot 1} = 121\ \Omega$ (compared to 130 Ω).

Lastly, N = (121 / 8)$^{0.5}$ = 3.89 (compared to 4.2).

Not a bad result, and given the wide range of BJT parameters we must deal with, this simplified procedure is quite acceptable.

With so little resistance in the collector circuit, the CE power amplifier above may have stability problems. For safety's sake, we must add emitter resistance. Using the stability factor S = 15, we can calculate a value of R$_E$.

$$S = \frac{1 + \dfrac{R_E}{R_B}}{\dfrac{1}{\beta + 1} + \dfrac{R_E}{R_B}} = 15$$

Then, assuming β = 150,

$$\frac{R_E}{R_B} = 0.241$$

Solving for R$_E$ gives R$_B$ = 64 Ω. We have $R_E = 0.241 \cdot 64 = 15\ \Omega$.

Efficiency drops slightly from 35% to 32%, but we know our design is stable.

After adding a suitably sized heat sink, we could quickly scale up our design to reach the 5-watt power output needed for our high-fidelity Amplifier. The problem is the size and cost of the audio transformer required. It would be custom-manufactured and expensive.

Recall that we needed a transformer to match the BJT's high output resistance to the 8 Ω speaker. Why not switch to a CC configuration that we know has a lower output resistance? We could easily make up the lower voltage gain in the driver stage.

The figure below shows a CC amplifier driving an 8 Ω speaker load directly in the emitter circuit.

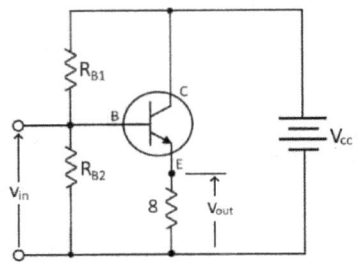

We can use a load line to evaluate the circuit's performance for our design. Below is a typical 40-watt power BJT collector characteristics with a $P_{cMax}$ = 30 w curve.

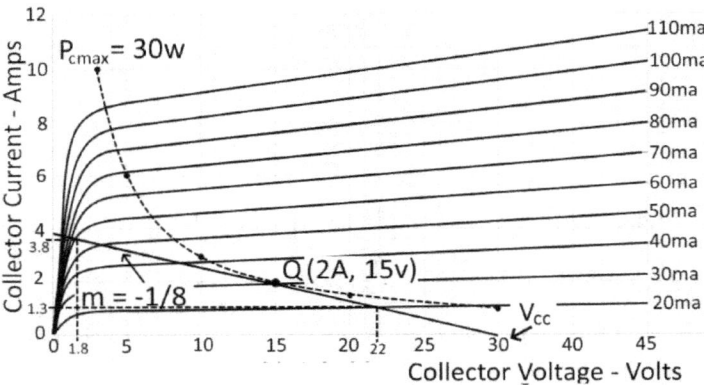

In designing our CC amplifier, we draw a load line tangent to the $P_{cMax}$ curve with a slope m = $-1/R_E$ = $-1/8$. While not ideal for maximum power output, it is the only load line with the required slope that does not cross the forbidden $P_{cMax}$ curve. If we moved it further up the $P_{cMax}$ curve, the load line would cross into forbidden territory.

We can calculate the power delivered to the 8 Ω load from the current and voltage excursions.

$$P_{out} = \frac{\Delta v \cdot \Delta i}{8} = \frac{20.2 \cdot 2.5}{8} = 6.3 \ w$$

This is not an efficient result given that we expend 60 w ($I_Q \cdot V_{cc}$) power to achieve 6.3 w output, but it meets our 5-watt specification without a costly output transformer.

Again, however, we have the problem of 2 A emitter current passing through our 8 Ω speaker! We can overcome this problem by introducing an emitter resistor and coupling capacitor to the speaker. See the modified circuit below.

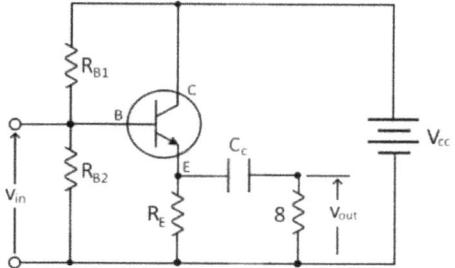

We must now select the value of $R_E$ using load line analysis.

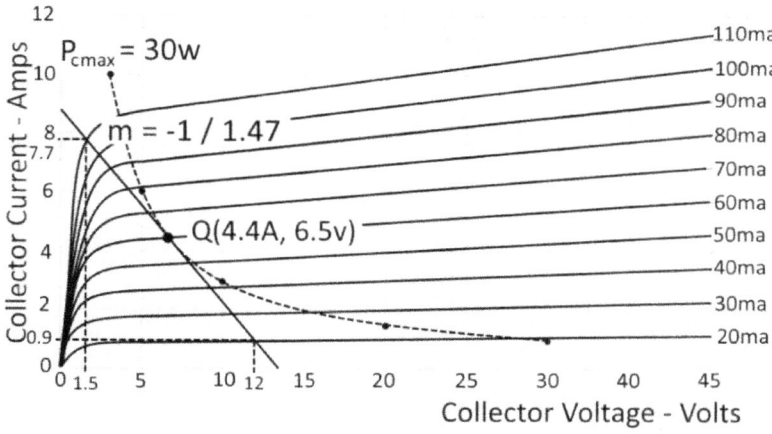

We draw an AC load line tangent to the 30-w curve, taking advantage of the maximum current and voltage change as shown. The slope of the AC load line is $m = -1 / (13.4v / 8.8 A) = -1 / 1.47$. Therefore, $R_E \mathbin{||} 8\,Ω = 1.47\,Ω$, makes $R_E = 1.8\,Ω$.

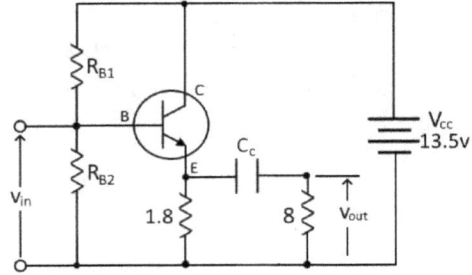

The power delivered to both resistors is,

$$P_{out} = \frac{\Delta v \cdot \Delta i}{8} = \frac{10.5 \cdot 6.8}{8} = 8.9 \ w$$

We waste $1.8 \cdot 4.4 = 7.5$ w power in $R_E$, with only 1.4 w delivered to the speaker! We cannot win for losing. Single-ended, Class A power amplifiers come up short again. There must be a better way.

And there is! In the next chapter, we explore push-pull amplifier configurations that neatly solve the power and efficiency problem and do so cost-effectively with transformers.

# Chapter 16 – Transformer Push-Pull Power Amplifiers

In the previous chapter, we configured single-ended BJT power amplifiers and found them to be inefficient and costly for high-power output designs. In this chapter, we explore push-pull BJTs configured as shown below.

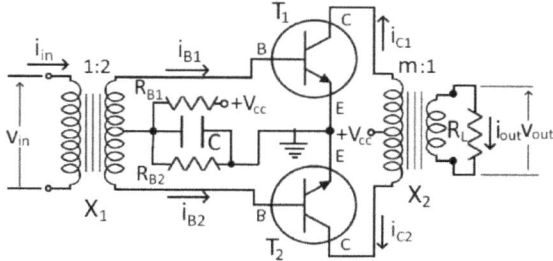

While we will eventually eliminate transformers, they provide an effective way to introduce push-pull BJT amplifiers. In addition, the circuit above is of historical value because it played a pivotal role in the design of the first BJT portable radios.

Looking at the circuit above, transformer $X_1$ drives the BJT bases with identical and inverted signals. For a sinusoid signal, "inverted" means one is "180 degrees out of phase" with the other. With this arrangement, $T_1$ is experiencing a current increase (a push) on the upswing of $v_{in}$, while $T_2$ is experiencing an identical current decrease (a pull) on the downswing of $v_{in}$. Thus, we arrive at the term *push-pull*. The output transformer combines the 180 degrees out of phase and amplified versions of $V_{in}$ and drives the load $R_L$ ( a speaker in a portable BJT radio). The output transformer also matches the output resistance of $T_1$ and $T_2$ to $R_L$.

In the circuit above, resistors $R_{B1}$ and $R_{B2}$ set the base bias and the quiescent point for the BJTs. Capacitor C short circuits the transformer center tap to common, as is necessary for a CE configuration.

To see the advantage of the push-pull arrangement, we will explore what happens to the input signal $v_{in}$ as it passes through the amplifier. When we apply $v_{in}$ across input transformer $X_1$, it sets up a current $i_{in}$ in $X_1$'s primary that splits in its secondary into currents $i_{B1}$ and $i_{B2}$ given by,

$i_{B1} = i_{in}$ and $i_{B2} = -i_{in}$

Output transformer $X_2$ combines the amplified signals such that,

$i_{out} = i_{C1} - i_{C2}$

Provided the BJTs are identical, the two output signal currents, $i_{C1}$, and $i_{C2}$, will be the same magnitude, except $i_{C2}$ will be the negative of $i_{C1}$. Thus, we have,

$i_{out} = i_{C1} - i_{C2} = A_i \cdot i_{B1} - (- \cdot i_{B2}) = A_i \cdot (i_{B1} + i_{B2}) = 2 \cdot A_i \cdot i_{in}$

where $A_i$ is the current gain of each BJT. The final output is double the output of a single-ended CE amplifier.

The doubling of gain is not a significant advantage in that we are using two BJFs. The real advantage accrues with the reduction of distortion. Issues with signal fidelity can be significant in BJT amplifiers operating at higher power output levels. The variation of $\beta$ over large collector current excursions mishaps the output waveform. The result is an output signal that, though amplified, is not an identical representation of the input signal. We measure the lack of signal fidelity by the amount of *harmonic distortion*.

To explore harmonic distortion, we will assume that the input signal to a push-pull amplifier is,

$i_{in} = \sin(\omega t)$

The corresponding output is

$i_{out} = I_c + I_0 + I_1 \sin(\omega t) + I_2 \sin(2\omega t) + I_3 \sin(3\omega t) + \cdots$

$I_c + I_0$ represents the DC offset current appearing at the output. The desired output signal is $I_1 \sin(\omega t)$ where $I_1$ is the magnitude of the amplified signal. The remaining terms constitute undesirable harmonic currents that misshape the output signal.

We compute total harmonic distortion (THD) this way.

$$THD = \frac{\sqrt{I_2{}^2 + I_3{}^2 + I_4{}^2 + \cdots}}{I_1} \cdot 100\%$$

As we show in Appendix D, the push-pull amplifier outputs are,

$i_{out1} = I_c + I_0 + I_1 \sin(\omega t) + I_2 \sin(2\omega t) + I_3 \sin(3\omega t) + \cdots$

$$i_{out2} = I_c + I_0 - AI_1\sin(\omega t) + I_2\sin(2\omega t) - i_3\sin(3\omega t) + \cdots$$

When passed to the output transformer,

$$i_{out} = i_{out1} - i_{out2} = 2(I_1\sin(\omega t) + I_3\sin(3\omega t) + \cdots)$$

The even-ordered harmonic terms have disappeared, reducing THD. So, too, have the constant terms, which leads to a second advantage: the transformer core does not experience the constant level of magnetic flux that exists with a single-ended amplifier. This means that watt-for-watt, push-pull output transformers are smaller and less expensive to construct.

To analyze the transformer push-pull amplifier, we will use load line analysis applied to *composite collector characteristics* like the one shown below for the 2N3904 BJT.

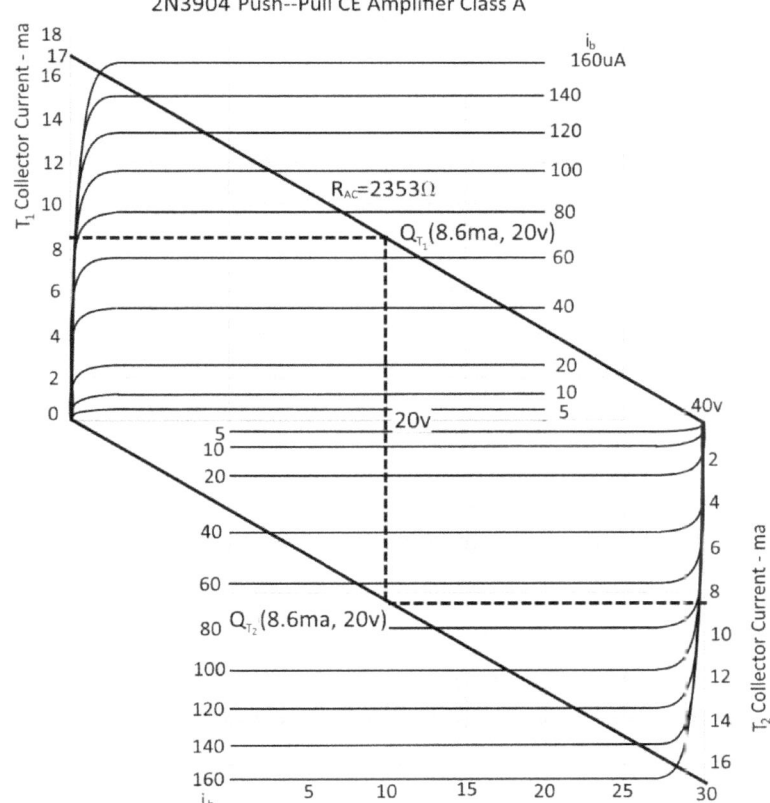

Note: The choice of 2N3904 for this example is quite reasonable. For a portable BJT radio, the power output is less than 200 milliwatts, which the 2N3904 can readily handle.

To make a composite collector characteristic, we attach an inverted and reversed 2N3904 collector characteristics to a normal one, aligning the $I_c = 0$ line for both with the $V_{cc} = 20$ volts point on the $V_{CE}$ axis. (We assumed $V_{cc} = 20$ volts for our example.) We then draw the AC load line for maximum output power from the 40-volt point on the $T_1$ voltage axis to the upper knee of the $T_1$ curves.

We calculate $R_{AC}$ as,

$$R_{AC} = \frac{40v}{0.017ma} = 2353\Omega$$

If m is the overall turns ratio for the push-pull output transformer and $R_L$ is the output (speaker) load, then,

$$R_{AC} = \frac{m^2}{2} \cdot R_L$$

Assuming an 8 Ω speaker, we solve for the turns ratio m and find that,

$$m = \sqrt{\frac{2 \cdot R'_L}{R_L}} = \sqrt{\frac{2 \cdot 2353}{8}} = 24.25$$

After finding values for $R_{B1}$ and $R_{B2}$, the final circuit is,

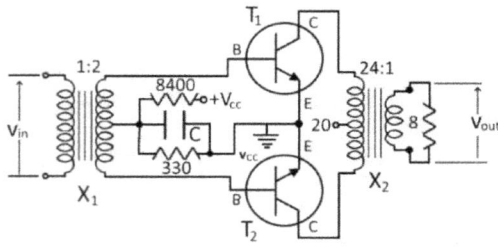

From the composite collector characteristic above, the maximum power output is,

$$P_{out} = \frac{38\,v \cdot 0.016\,ma}{8} \cdot 2 = 152\,mw$$

Here are the results from the simulation.

$P_{out}$ (before clipping) $= 160$ mw in the 8 Ohm load

To check the cancelation of even-ordered harmonics, we measured 12% second-order harmonic distortion in the collector current of $T_1$. In the 8 $\Omega$ load, the second-order distortion dropped below 1%.

Total power consumption was about (20 v · 20 ma) 400 mw, so the efficiency was about 38%. The 150-mw output would suit a portable BJT radio, but the continuous power consumption is too great.

Happily, the push-pull configuration offers a clever way to reduce power consumption. Before detailing a solution, however, it is helpful to define the *operating classes* of push-pull amplifiers. In *Class A operation*, both output BJTs are in complete turn-on during the entire signal cycle. Their collector quiescent point is such that they consume power even when no signal is present. While Class A operation is the least efficient, it produces the least THD. The push-pull amplifier above operates Class A.

In *Class B operation*, we bias the output BJTs just below their turn-on current. The result is that each BJT operates only during half of the signal's cycle. For an NPN BJT pair, this is the positive half-cycle; for a PNP BJT pair, it is the negative half-cycle. The output transformer combines the two half cycles to form the complete signal. In the absence of a signal, only cutoff current flows.

Early BJT portable radios used a Class B push-pull output stage with two miniature and inexpensive audio transformers. (Of interest, the output BJTs were two germanium PNP BJTs.) The design provided suitable speaker volume, and its low power consumption offered a reasonable battery life of 20+ hours.

A downside of Class B operation is that it produces an irregularity in the amplified signal near the crossover point between the turn-on voltage of the BJTs. We call this *crossover distortion.* Crossover distortion is most apparent at low speaker volume. As a practical matter, Class B push-pull amplifiers traded power consumption for distortion, which was acceptable in early portable BJT radios.

A way to reduce crossover distortion involves biasing the output transistors to a point between Class A and Class B, which we designate *Class AB operation*. In Class AB, we bias each collector current to just beyond turn-on. The result is that each BJT amplifies slightly more than half of the input signal. As before, the output transformer combines the two out-of-phase signals to form a complete signal, now without an irregularity at the crossover point. The Class AB's operational current overlap reduces crossover distortion while retaining Class B's power-saving advantages.

The base biasing decision in Class AB operation is a tradeoff. The closer we come to full Class A, the less distortion, but the greater the power consumption. Typically, we choose for Class AB quiescent current about 10-15% of Class A quiescent current.

To explore Class B and AB operation with simulation, we modified the circuit above by replacing $R_{B1}$ with a variable resistor. We can adjust $R_{B1}$ to simulate any class of operation.

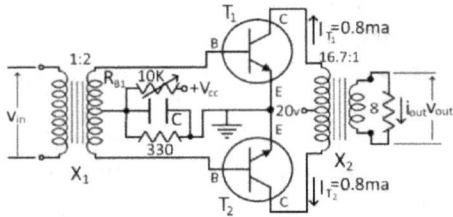

We first adjusted $R_{B1}$ for Class B operation. We set the BJTs' collector current to 0.8 ma (10% of original $I_Q$ = 8 ma), slightly below $V_{BE}$ turn-on potential. A positive going signal overcomes the remaining turn-on potential of $V_{BE}$ and moves the collector current into its normal operational range. $T_1$ will amplify slightly less than one-half of the input signal, and $T_2$ will amplify slightly less than the other half.

We can use the composite collector characteristics to predict how the amplifier will function.

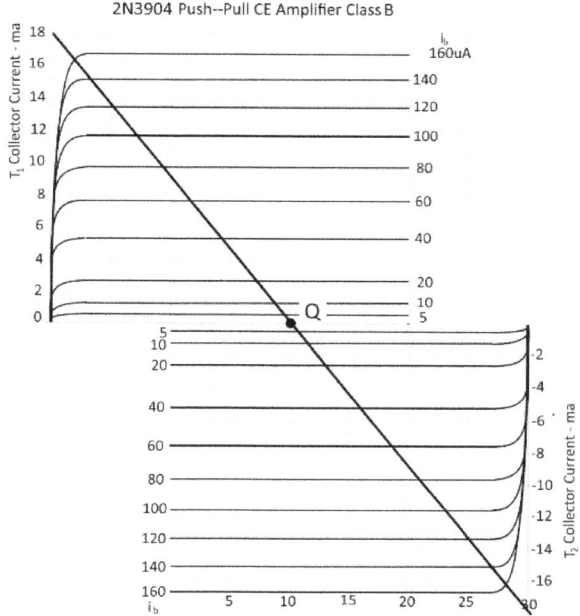

The quiescent point for both BJTs is on the $I_c = 0$ axis, as shown. The load line of maximum output is between the knees of the $i_c$ curves, which the figure shows. The slope of the load line is,

$$R_{AC} = \frac{20v}{0.018A} = 1111 \ \Omega \ \text{ and } m = \sqrt{\frac{2 \cdot R'_L}{R_L}} = \sqrt{\frac{2 \cdot 1111}{8}} = 16.66$$

Here are the simulated results.

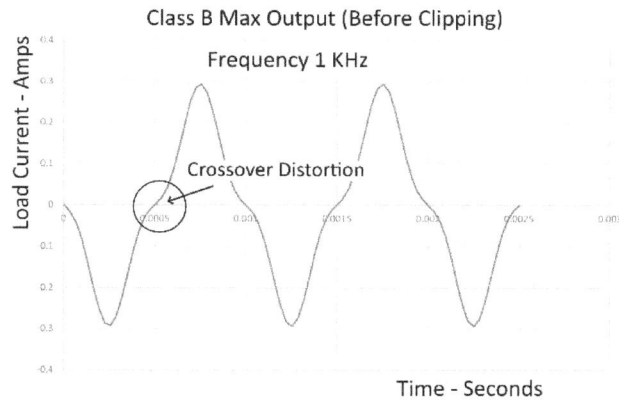

The crossover non-linearity is quite apparent. The second-order harmonic distortion was slight (less than 1 %), but the third-order harmonic distortion was 24%. The power output was 240 mw while consuming only 424 mw for an efficiency of 57%. Despite the crossover distortion, the high efficiency made this a viable option for BJT portable radios.

To move to Class AB operation, we adjusted $R_{B1}$ to move the BJT quiescent points into the BJTs' operating range at $I_c$ =2 ma. See the composite collector characteristics below.

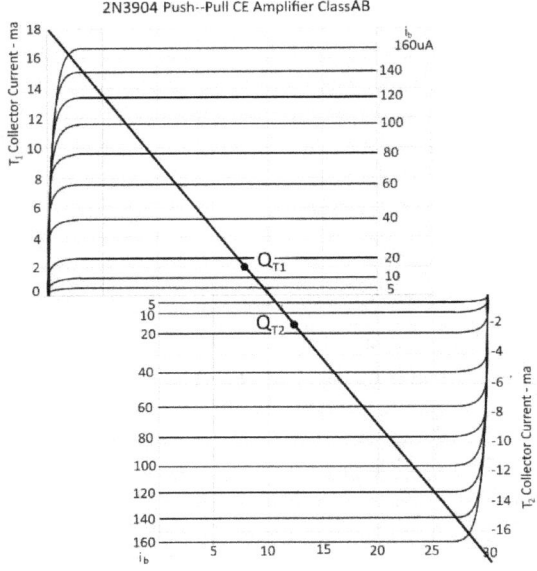

The load line is the same as the Class B amplifier, with only the quiescent points shifted into the operating range. T1 and T2 amplify more than 50% of the input signal, removing the crossover distortion. Here is the resulting output.

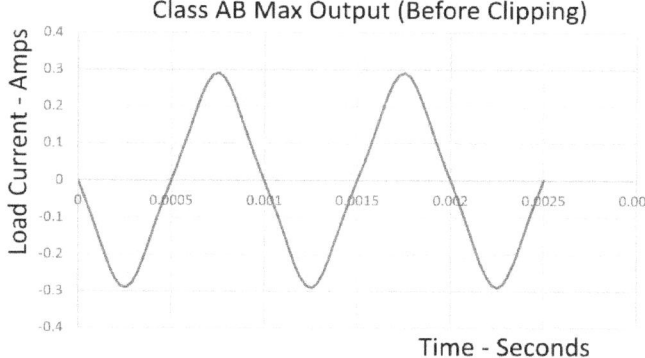

**Class AB Max Output (Before Clipping)**

Crossover non-linearity is hardly noticeable. With the much reduced third-order distortion, THD was an acceptable 9%. Efficiency decreased slightly to 51%, still a significant improvement on Class A of 35%.

Note: The author recalls early high-fidelity BJT amplifiers that used Class AB with bulky and expensive transformers. Obviously, these amplifiers were too expensive to compete with tube amplifiers at the time. But, as always, technology does not stand still, and competition drives the market to find a solution.

While push-pull Class AB operation is a good option for our high-fidelity amplifier, we still must eliminate costly transformers. In the next chapter, we offer a transformerless solution!

# Chapter 17 – Direct Drive Push-Pull Power Amplifiers

In Chapter 15, we learned that CC amplifiers, with their low-output resistance, made them good prospects for power output designs. The problem for single-ended CC power amplifiers was (1) we could not use a speaker for the load as the DC current would disrupt the speaker's operation and (2) the alternative, capacitive coupling, wasted too much power in the required collector resistor.

With BJT technical advances, power amplifier designers soon found a way to overcome these CC design limitations. The advancement came with the development of *complementary BJTs,* NPN-PNP pairs with the same collector characteristic differing only in the polarity of operating voltage.

As an example, the 2N3904 NPN BJT has a PNP complement, the 2N3906. An amplifier circuit using a 2N3904 works equally well with a 2N3906 by reversing the polarity of the supply voltage. In this chapter, we will use the TIP41-TIP42 complementary pair as the CC output stage of our high-fidelity amplifier design.

## Class B Push-Pull

To explore complementary BJT amplifier design, we will begin with the bare bones complementary push-pull circuit below.

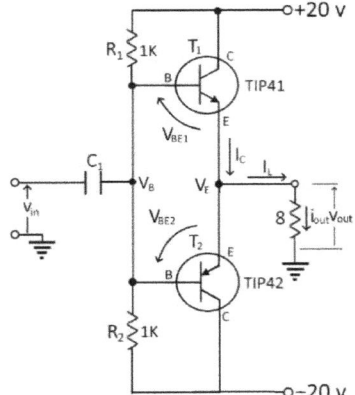

We set the BJT's base bias for Class B operation. If we match the BJTs closely and $R_1$ and $R_2$ are equal, the emitter voltage $V_E$ and base voltage $V_B$ will be zero. With $V_{BE} = 0$, both BJTs will be in cut off ($I_C = 0$). Also, since $V_E$ is zero, no current will flow in the 8 $\Omega$ load. ($I_L$ being zero is a necessary condition if the load is to be speaker.) Under these conditions, the base current for the TIP41 and TIP42 is small, less than 1 ma each or 2 ma total. We want the current in $R_1$ and $R_2$ to be at least 10 times this, so we make it 10 · 2 ma = 20 ma. Thus, we choose $R_1$ and $R_2$ each to be 20 v /0.002 A = 1000 Ω.

If we apply a sinusoidal signal $v_{in}$ to the TIP41 and TIP42 bases, the following cycle takes place:

1. On the signal's positive half cycle, $V_{BE1}$ rises above $T_1$'s cutoff voltage, and $T_1$ operates as a CC amplifier driving the 8 Ω load.
2. On the signal's negative half cycle, $T_1$ cuts off, $V_{BE2}$ rises (in a negative sense) above $T_2$'s cutoff voltage, and $T_2$ operates as a CC amplifier driving the 8 Ω load.
3. Eventually, the signal returns to zero, $T_2$ cuts off, and the cycle is set to repeat.

In short, $T_1$ supplies the 8 Ω load's signal current during the signal's positive half cycle, while $T_2$ does the same during the negative half cycle. Taken together, the entire input signal appears in the 8 Ω load.

Since we are dealing with a CC amplifier, the voltage gain is one. We can readily make up for any needed voltage gain in the driver stage.

We simulated the Class B circuit above with this result. At about 1 watt output (1 KHz sinusoidal signal), even-order distortion components were unmeasurable, while odd-order measured about 8%.

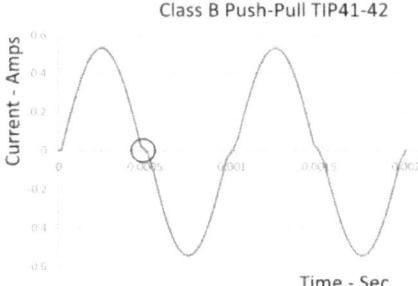

Class B Push-Pull TIP41-42

At low signal levels, crossover irregularity was more prominent, and distortion increased significantly.

## Class AB Push-Pull

Our next step was to switch from Class B to Class AB operation. The simplest way to do this was to introduce a resistor of suitable value between $R_1$ and $R_2$.

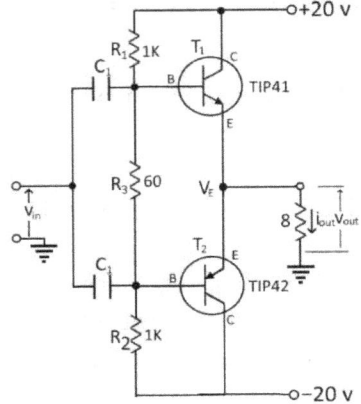

The presence of $R_3$ increased the BJTs' base-emitter voltage ($V_{BE1}$ became more positive and $V_{BE2}$ more negative), moving the BJTs in the direction of turn-on. In the simulation, we adjusted $R_3$ for a collector current of 100 ma, about 10% of the Class A value of 1 A. As expected, the crossover distortion disappeared.

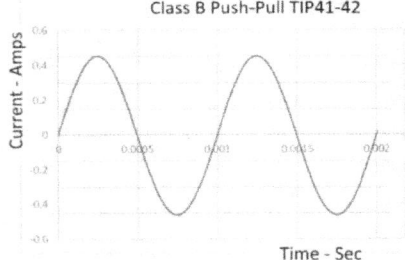

THD (both even and odd order) dropped to less than 1%. In addition, we were able to reach our target goal of 5 w without clipping.

Before breadboarding, we needed to protect our circuit against thermal instability. This included (1) the addition of bias diodes, (2) the addition of emitter resistors, and (3) base clamping BJTs.

## Diode Thermal Stabilization

The circuit above does not address thermal stabilization. One strategy to do so involves replacing $R_3$ with two forward-biased silicon diodes, which together would produce the 1.2 v differential needed between BJT bases to approach turn-on. We would physically attach the diodes to the BJTs so they would experience the same heating. Then, provided the thermal characteristics of the diodes and BJTs were well matched, the diodes' would change in such a way as to compensate for the rise of heat and thereby deter thermal runaway.

In the simulation, two diodes produced too much differential (nearer 1.4 v) for Class AB operation. We eventually found that one diode and a suitably sized resistor worked best. We estimated the resistor's value this way. Assuming the diode voltage was 0.6 v, we needed a resistor value equal to 0.6 / 0.02 or 30 Ω. As this value was highly dependent on the TIP41 and TIP42's $V_{BE}$ characteristics, our final design used a 100 Ω trim pot in parallel with a 50 Ω resistor. See below.

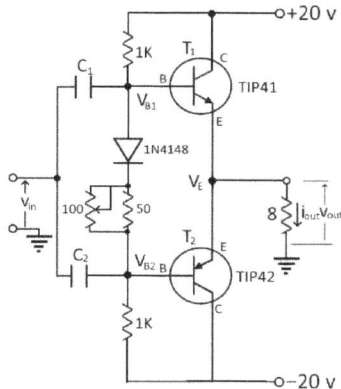

## Emitter Resistor Thermal Stabilization

Not being confident that one diode would provide sufficient thermal stability, we added a stabilizing resistor in the emitter circuit of each BJT. To determine the resistor value, we reasoned this way. Suppose the BJTs undergo a 40 °C rise in temperature. Base-emitter voltage would rise about 2 mv per °C, so we would expect $V_{BE}$ to increase by 40 · 2 = 80 mv. Consulting the TIP41-TIP42 data sheets, we found the relation between $V_{BE}$ and $I_c$ shown below.

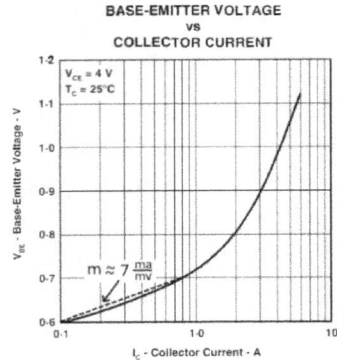

In our operating $I_c$ range 0.1 A to 1 A, collector current rose about 7 $\frac{ma}{mv}$. For a 40 °C rise in temperature, the increase in collector current would be $\Delta I_c = 7 \cdot 80$ = 560 ma. For the drop across the emitter resistor to be at least equal to 80 mv, we found that $R_{Emin} = \frac{80}{560} = 0.14\ \Omega$. We elected to use a commonly available 0.33 Ω resistor, which would provide even better thermal stability.

Here is the circuit with the 0.33 Ω emitter resistors in place.

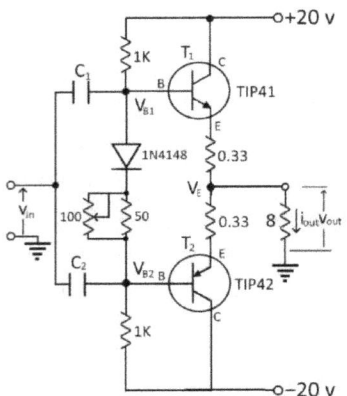

Note: Adding emitter resistors decreased efficiency slightly, but the added thermal stability was well worth it.

## Base Clamping Stabilization

While the combination of diode and emitter resistors was more than adequate, we decided to go a step further. As shown below, we added two clamping BJTs to the bases of the output BJTs.

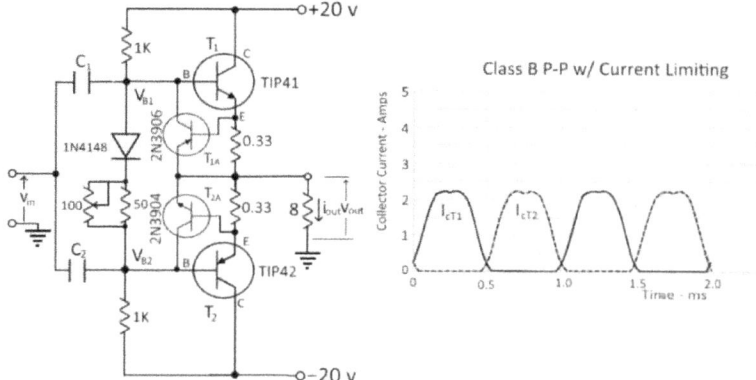

When the drop across each emitter resistor reached ~0.7 v, the BJTs turned on and prevented further increase in the output BJTs' base-emitter voltage. This clamped the BJT collector current at about 0.7 / 0.33 = 2 A and produced a predictable and more certain control of maximum collector current.

We simulated the final circuit above and found the results excellent. It could readily produce 5 w audio with less than 1% THD.

We have now moved to the question of overall voltage gain. To obtain 5 w of audio across the 8 Ω speaker load, we calculated that we needed an ~18 vpp signal.

$$\frac{(^{18}/_{2 \cdot \sqrt{2}})^2}{8} = 5.06w$$

The input specification called for 5 w to be produced with an input of 0.5 vpp. With the CC output stage voltage gain one, we needed a driver stage with a voltage gain of at least 36. This is readily achievable with a CE driver stage.

We used a direct drive CE stage, eliminating capacitors $C_1$ and $C_2$. Our reasoning went like this. We would use a CE BJT's collector-emitter resistance as a resistor equivalent of the lower 1000 Ω base bias resistor. In this way, we could vary the driver BJT's base bias to control the TIP41 and TIP42's base bias and zero $V_{out}$.

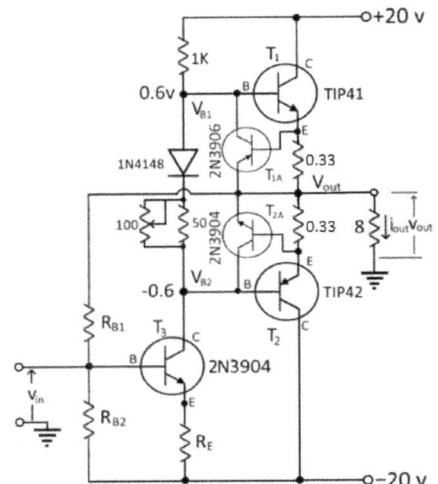

By varying $T_3$'s base bias, we controlled $V_{B2}$ and the DC output voltage $V_{out}$. Also, by making $R_{B1}$'s voltage source $V_{out}$, we introduced negative DC and AC feedback. Once we set $V_{out}$'s DC level to zero, changes in $V_{out}$ would feed back to $T_3$'s base and provide a corrective change in $T_3$'s collector current. This, in turn, would bring $V_{out}$ back to zero. While AC negative feedback reduced gain, it widened frequency response and decreased harmonic distortion even further.

We needed to find values for $R_{B1}$ and $R_{B2}$ that would set the DC output level to 0 v. We first assumed that $R_B = 1800\ \Omega$. We then selected the value of the emitter resistor $R_E$ so that $T_3$'s stability factor would be 25 or less.

$$S = \frac{R_B + R_E}{\dfrac{R_B}{\beta + 1} + R_E} = \frac{1800 + R_E}{\dfrac{1800}{150 + 1} + R_E} \leq 25$$

Solving, we found that,

$$R_E \geq 62\ \Omega$$

We chose $R_E = 68\ \Omega$

From the circuit above, we saw that the current in $R_E$ was,

$$I_{R_E} = \frac{20 - 0.6}{1000} = 19.4\ \text{ma}$$

The voltage drop across $R_E$ was then 68 · 0.0194 = 1.32 v. If we wanted $V_{BE}$ for $T_3$ to be 0.7 v, then $V_B$ had to be,

$$V_B = 0.7 + 1.32 = 2.02v.$$

We now had two equations in two unknowns.

$$R_B = \frac{R_{B1} \cdot R_{B2}}{R_{B1} + R_{B2}} = 1800 \text{ and } 20 \cdot \frac{R_{B2}}{R_{B1} + R_{B2}} = 2.02$$

Solving, we found that,

$$R_{B1} = \frac{1800}{0.101} = 17821 \ \Omega \approx 18K \ \Omega \text{ and } R_5 = 2000 \ \Omega \approx 2.2K$$

Rather than using a fixed 18 K Ω resistor for $R_{B1}$, we used a 10 KΩ trim pot in series with a 10 KΩ resistor.

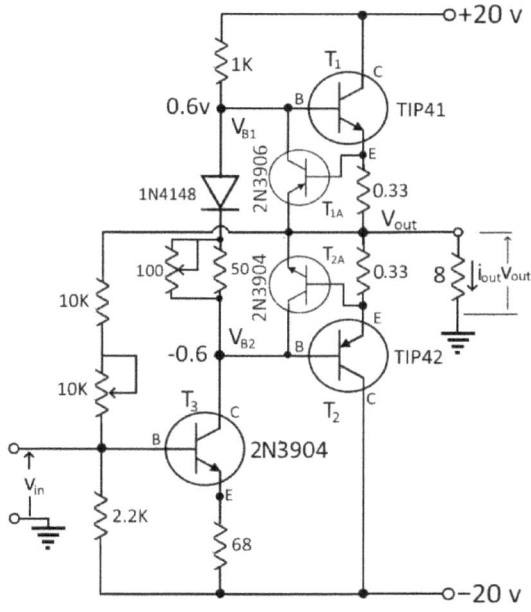

During simulation, we used both trim pots to zero $V_{out}$ and bring the collect current of TIP41 and TIP42 to ~100 ma. With 1.4 vpp input, we measured 3 w output into the 8 Ω load. Supply power was ~7 w, giving an efficiency of 43%. The maximum power out before clipping exceeded 6 w.

In this chapter, we relied exclusively on simulation to test the power amplifier stages. Before moving on to breadboarding, we needed to select a suitably sized heat sink for the TIP41 and TIP42. We focus on heat sink selection in the next chapter.

# Chapter 18 – Heat Sink Design

Before delving into this subject, here are "must know and heed" items when working with BJT power BJTs.

1. <u>Never</u> operate a power BJT without a suitably sized heat sink. Even a short signal burst can initiate a thermal runaway without a heat sink.
2. Most power BJTs electrically interconnect the collector and case. Never mount more than one BJT on the same heat sink unless electrically insulated.
3. A too-large heat sink is preferable, as power BJTs work best when cool.

Now that we have that out of the way, we start with fact number one: The temperature of the power BJT's junction $T_j$ is critical to the BJT's performance. In the author's experience, never allow the temperature to exceed ambient (25°C) by more than 60°C; that is, use Tj = 85°C to size a heat sink for a power BJT.

An effective way to think of how to cool a power BJT is to use an analogy to electrical resistance. Here are the elements of the analogy:

| Electrical | Heat |
| --- | --- |
| Electrical Resistance R (Ω) | Thermal Resistance Θ (°C/Watt or °C/W) |
| Voltage V (volts) | Temperature T (°C) |
| Current I (amps) | Heat Flow P (Watts or W) |

Recall that we measure current in amps, which is coulombs of electrons flowing per second. We measure heat flow in watts, which is Joules of thermal energy flowing per second. The voltage difference across a resistor causes current (electrons) to flow. Similarly, temperature difference across a thermal resistance causes heat energy to flow. In equation form, we have,

$$P = \frac{T_H - T_L}{\theta} = \frac{\Delta T}{\theta}$$ or rewritten $\Delta T = \theta \cdot P$, which is analogous to Ohms Law $\Delta E = R \cdot I$.

Also, in electrical circuits, we say that current flows from a point of higher voltage to a lower one. Similarly, in thermal systems, we say that heat flows from a point of higher temperature to a point of lower temperature.

The figure below shows a power BJT attached to a heat sink.

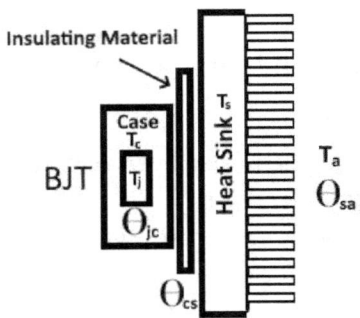

The Ts in the figure are the temperatures of the different elements. We begin with BJT junction $T_j$ and conclude with the temperature of the surrounding air $T_a$. Associated with each element is its thermal resistance, beginning with the BJT junction to case value $\Theta_{jc}$, followed by the case to heatsink value $\Theta_{cs}$, and lastly, the heat sink to air value $\Theta_{sa}$.

Heat flows from the BJT junction through each thermal layer, eventually dissipating in the surrounding air. We deduce from $\Delta T = \theta \cdot P$ that the difference in junction-to-air temperature is directly proportional to the thermal resistance of all the intervening thermal elements. Therefore, if the heat flow P encounters too large a total thermal resistance $\Theta_{Total}$, the junction's temperature will exceed the recommended BJT's maximum, and damage is likely to occur.

How do we calculate the total thermal resistance $\Theta_{Total}$ between the BJT junction and the surrounding air? As it turns out, the thermal layering acts like a series resistance in an electrical circuit, meaning that,

$$\theta_{Total} = \theta_{jc} + \theta_{cs} + \theta_{sa}$$

$\Theta_{jc}$ is the thermal resistance of the BJT's junction to the case. We usually find its value on the BJT's data sheet. For example, the TIP41-42 data sheet shows two values:

$$\theta_{jc} = 1.92 \; \frac{°C}{W} \text{ and } \theta_{ja} = 62.5 \; \frac{°C}{W}$$

In an actual amplifier, we would never use the thermal resistance to surrounding air the second value). Why? Because we always use a heat sink!

$\Theta_{cs}$ is the case to heat sink thermal resistance. When we mount the BJT directly to the heat sink, $\Theta_{cs}$ is about 0.6 °C/W. If we use a thermal grease, we reduce the value to 0.15 °C/W. If we need to insulate the BJT from the heat sink, we insert a mica or other specially-made insulator. This typically increases $\Theta_{cs}$ to 1.2 to 1.6 °C/W. Adding thermal grease, we reduce $\Theta_{cs}$ to 0.3 to 0.4 °C/W.

$\Theta_{sa}$ is the heat sink to surrounding air thermal resistance. The heat sink manufacturer specifies this value. $\Theta_{sa}$ varies widely depending on the heat sink's size and shape.

To specify a heat sink, we start by estimating the minimum thermal resistance needed for a specified power dissipation, then search out a heat sink with at least this value.

For our Class AB amplifier of Chapter 17, we will use the 20% rule developed in Appendix E. The rule states that each BJT's maximum dissipation is 20% of the amplifier's maximum output power plus the quiescent power dissipated ($I_Q \cdot V_{CEQ}$). We calculate the maximum output power using the maximum peak-to-peak value of $v_{out}$. So, we used $V_{cc}$ = 20 volts in the calculation below.

$$P_{out\ max} = \frac{v_{rms}^2}{R_L}$$

$$v_{rms} = \frac{V_{cc}}{2 \cdot \sqrt{2}} = \frac{20}{2.828} = 7.07\ vrms$$

$$P_{out\ max} = \frac{7.07^2}{8} = 6.25\ watts$$

$$P_{D\ T_1} = 0.20 \cdot 6.25 + 20 \cdot 0.1 = 1.25 + 2.0 = 3.25\ watts$$

So, the maximum power dissipation in our heat sink calculation was 3.25 watts.

$$\theta_{Total} = \frac{(85 - 25)°C}{3.25\ w} = 18.5\ \frac{°C}{w}$$

$$\theta_{Total} = 1.92 + 0.15 + \theta_{sa} = 20.6\ \frac{°C}{w}$$

$\theta_{sa} \approx 20.6\ \frac{°C}{w}$, which was the minimum thermal resistance of the heat sink we needed.

We selected a possible TO-220 heat sink from those available. The heat sink's specification graph showed two plots.

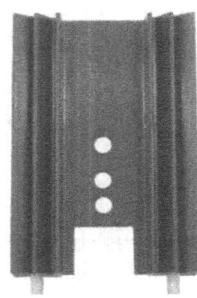
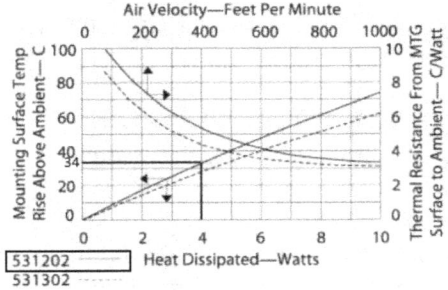

The plot with axes at the top and right is for a heat sink force-air cooled. The axes on the left and bottom are for a vertically mounted heat sink with only convection air flow for cooling. The latter is how we planned to use the heat sink, so we used that plot. We further assumed the plot was a straight line over the $P_c$ range shown on the graph. Given this assumption, the slope of the line was $\frac{1}{\theta_{sa}}$ and, therefore, $\theta_{sa} = \frac{75}{10} = 7.5 \, \frac{°C}{W}$.

Since $7.5 \, \frac{°C}{W} < 20.6 \, \frac{°C}{w}$, the chosen heat sink was more than adequate for our power BJT.

In computing maximum output, we used $V_{cc}$ as the maximum peak-to-peak value of the output voltage. The output voltage in actual amplifiers will not reach this value, so using Vcc built-in additional headroom for our heat sink calculation. How much was the headroom? Our simulated TIP41-42 amplifier had a maximum $v_{out}$ of 14 vpp, less than the 20 vpp we used in the calculation. Thus, the selected heat sink was more than adequate.

We completed the power amplifier design with the selected heat sink. In the next chapter, we add a preamp stage.

# Chapter 19 – Putting It All Together

Satisfied with the simulated design of Chapter 17 and having mounted the TIP41 and TIP42 on an appropriately sized heat sink, we were ready to breadboard test the driver and power output stages.

Before applying power, we set the base bias 100 Ω trim pot to 0 Ω to place the TIP41 and TIP42 in Class B operation. We then adjusted the 10 KΩ trim pot to 8 KΩ to set 18 KΩ as a starting value for $R_{B1}$.

With $V_{cc}$ set to ±5 v, we powered up and read a satisfactcry 12 ma total current, which, with the TIP41 and Tip-42 in Class B, was the driver BJT current alone. Next, we adjusted the 10 KΩ trim pot to place the output voltage $V_{out}$ at zero. Then, we slowly increased the 100 Ω trim pot value until the total current was 50 ma. The rise in total current indicated that the TIP41 and TIP42 BJTs were advancing toward turn-on.

To be cautious, we made only incremental changes in the supply voltages. Specifically, we raised the supply voltages ±5 volts at a time while (1) adjusting the 10 KΩ trim pot to keep $V_{out}$ at zero volts and (3) adjusting the 100 Ω trim pot to keep the supply current under 120 ma (our target Class AB value plus 20 ma driver current). We continued this procedure until we reached $V_{cc}$ = ±20 v. At that point, the supply current was a satisfactory ~120 ma, and $V_{out}$ was less than 20 millivolts.

Next, we applied an input signal and monitored output across the 8 Ω load. The output signal was clean and well-shaped at just shy of clipping. An input of 1.7 vpp produced 4 w audio with less than 1% harmonic distortion. The frequency response was down 0.5 DB at 20 Hz and flat well beyond 20,000 Hz.

The current design failed to meet the input level and 5-watt design specification. The quick solution was to increase ±$V_{cc}$ from ±20 v to ±30 v. After making this change, we readjusted both the driver 10 KΩ pot for $V_{out}$ = zero and 100 Ω trim pot for a total current of 120 ma. After this change, the amplifier reached 6 watts before clipping. The input voltage for 5 watts was 1.4 vpp, greater than the 1 vpp specification. We met the 5-W specification but were still shy of the input level. We needed 1.0 vpp to produce the 5 w.

In addition, the input resistance was ~2.2 KΩ, and we needed it to be 100 KΩ. To address this, we used the Darlington pair pre-amplifie⁻ of Chapter 14. With it, we could readily match the 100 KΩ input specification with the ~2.2 KΩ driver stage input resistance of the driver stage.

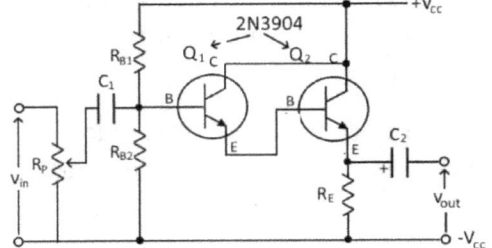

We also added a volume control on the front end. To achieve a minimum 100 KΩ input resistance, we used a 250 KΩ audio taper volume control, and then, by making $R_{B1} = R_{B2} = 470$ KΩ, the overall parallel combination of $R_{B1}$, $R_{B2}$, and $R_P$ was 121 KΩ minimum when the volume control was fully clockwise. For lesser volume settings, the input resistance would be greater than 121 KΩ. To complete the Darlington pre-amplifier, we chose 5 KΩ for $R_E$ to keep the $Q_2$'s collector current in the 4 to 8 ma range.

The low output resistance of $Q_2$, in conjunction with the low input resistance of the power amplifier driver, meant that the output coupling capacitor $C_1$ must be large, say, 50 μF. Given that value,

$$f_{low} \approx \frac{1}{2\pi \cdot 3200 \cdot 50 \cdot 10^{-6}} = 0.99 \text{ Hz.}$$

As for $C_2$, the input coupling capacitor, the opposite was true. A much smaller 0.1 μF sufficed.

$$f_{low} \approx \frac{1}{2\pi \cdot 500 \cdot 10^3 \cdot 0.1 \cdot 10^{-6}} = 3.0 \text{ Hz}$$

These values kept the low-frequency response within specification.

Here, then, is the final circuit.

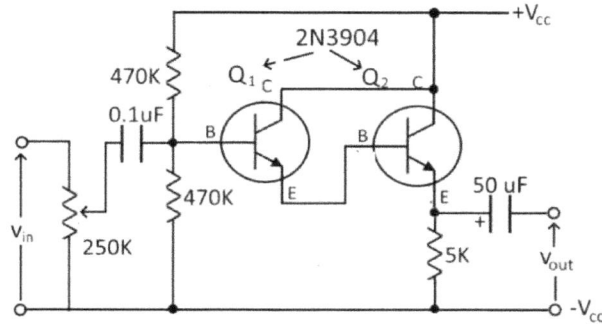

It was time to address the need for an increase in gain. Since the Darlington voltage gain was ~1, we had to look elsewhere. The place was the 68 Ω emitter resistor in the driver stage. Using simulation, we found that bypassing it entirely created too much gain. By trial and error (which we can easily do with simulation), we determined that a 250 µF capacitor in series with an 82 Ω resistor worked perfectly.

Putting the Darlington and power stages together, we had our final design.

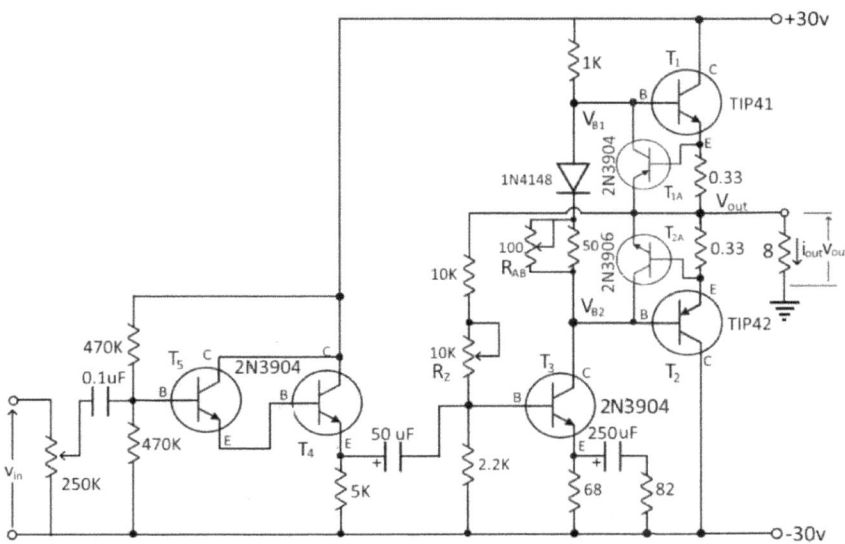

After breadboarding the circuit, here are the results.

1. Input sensitivity met: 1-volt peak-to-peak sinusoid of frequency 1000 Hz produced 5.2 watts output into an 8 Ω speaker load. (Maximum power output

before clipping was 6.4 watts with an input voltage of 1.3 vpp.)

2. Input load met: ~120 KΩ at full volume.

3. Frequency response met: Down 0.8 DB at 20 Hz. Flat passed 20,000 Hz. Down 1 DB at 100 KHz.

4. Harmonic distortion at 1000 Hz met: Less than 0.5%.

While the results were excellent with $V_{cc}$ = ±30 v, we experimented with a $V_{cc}$ = ±20 volts, as initially planned. Recall that the simulation indicated we could only reach the 5-watt output without clipping by raising $V_{cc}$ to ±30 volts. When we breadboarded the circuit with $V_{cc}$ = ±20, the amplifier quickly reached 5 watts output with 1 vpp input without clipping. The only difference was that total harmonic distortion rose from less than 0.5% ($V_{cc}$ = ±30 v) to about 0.9% ($V_{cc}$ = ±20 v), still within specification.

The lesson we learned (which we knew already) was that models like SPICE are imperfect. **Simulations have their use, but breadboarding is essential to pin down a circuit's actual operation and performance.**

# Chapter 20- Conclusion

We achieved our objective. Assuming no prior knowledge of binary junction transistors, we started with basic ideas, built on them, and finally applied our knowledge to design a BJT high-fidelity amplifier.

Whether you are inclined to design and build a BJT amplifier or not, we hope this book has provided good insight into basic semiconductor theory and BJTs. Moreover, other semiconductors and applications lie in wait for you to discover. Rest assured; the basic semiconductor theory introduced here will give you a head start if you pursue the subject further!

# Appendix A – Definitions

1. eV (electron-volt) – A unit of energy equal to $1.602176634 \times 10^{-19}$ Joules. While Joules is the regular metric unit, we more conveniently represent the exceedingly small size of atomic level measurements by the electron volt.

2. Symbol "a||b" – Indicates the parallel combination of components a and b. If a and b are resistance values, then,

$$a||b = \frac{a \cdot b}{a + b}$$

If a and b are capacitors, then,

$$a||b = a + b$$

3. Breadboard – Temporarily wiring an electronic prototype for test purposes. The name is derived from the early practice of using a wooden bread cutting board as the base for wiring the prototype.

# Appendix B – Stability Factor Design Example

A valuable use of S is as a basis of bias design. In this case, we pick the value of S and use a rearranged version of Eqn. 7-10 to find the ratio $R_E$ to $R_B$.

$$S = \frac{(1 + \frac{R_E}{R_B})}{\frac{1}{(\beta + 1)} + \frac{R_E}{R_B}} \quad (Eqn.\, B - 1)$$

With this ratio, we would proceed similarly to the design of Chapter 5.

As an example, design a CB bias scheme using these values:

$V_S$ = 20 volts
$V_{CE}$ = 10 volts
$V_{BE}$ = 0.7 volts
$R_B$ = 10 KΩ
$I_C$ (nominal) = 3.7 ma
$I_{SO}$ nA at 25° C
β (nominal) = 200
S = 15

If we solve Eqn. B-1 for $R_E$ / $R_B$, we obtain,

$$\frac{R_E}{R_B} = \frac{\frac{S}{\beta + 1} - 1}{1 - S} \quad (Eqn.\, B - 2)$$

Substituting S and β into Eqn. B-2, we calculate $R_E$ / $R_B$,

$$\frac{R_E}{R_B} = \frac{\frac{15}{200 + 1} - 1}{1 - 15} = 0.066$$

Since we specified $R_B$ as 10 KΩ, $R_E$ must be 660 Ω. Knowing $V_{CE}$, $V_S$ and now $V_E$ = 660 · 0.0037 = 2.44 volts. We calculate $R_C$ from,

$$R_C = \frac{20 - 10 - 2.44}{0.0037} = 2043 \,\Omega$$

To find $R_1$ and $R_2$, we first must find $I_B$. We start with Eqn. 7-3,

$$I_C = \beta \cdot I_B + I_{CEO}$$

Then note that $I_{CEO} = (\beta + 1) \cdot I_{SO} = 201 \cdot 50 \cdot 10^{-9} = 0.1$ ma. Solving for $I_B$ above,

$$I_B = \frac{.0037 - .0001}{200} = 18\ \mu A$$

If we assume that $I_E = I_C$, then

$$E_B = R_B \cdot I_B + V_{BE} + R_C \cdot I_C = 10000 \cdot 18 \cdot 10^{-6} + 0.7 + 660 \cdot 0.0037$$
$$= 3.32 \text{ volts}$$

Referring to Eqns. 6-12a and b,

$$R_1 = \frac{V_S \cdot R_B}{V_B} = \frac{20 \cdot 10000}{3.32} = 60{,}240\ \Omega$$

$$R_2 = \frac{R_1 \cdot R_B}{R_1 - R_B} = \frac{60240 \cdot 10000}{60240 - 10000} = 11990\ \Omega$$

The figure below shows the CE circuit with the calculated parameters.

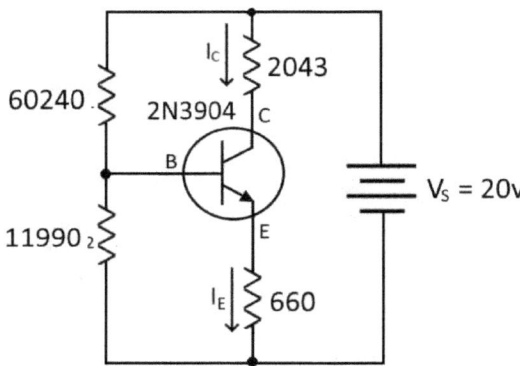

As a practical matter, we would need to select the nearest standard resistor values. Doing so could shift the quiescent point and require minor calculated value adjustments. For our purposes, we shall keep the calculated values.

# Appendix C – Maximum Power Transfer

To study power transfer, assume we have a voltage source $v_s$ with output resistance driving a load resistance $R_L$. See the circuit below.

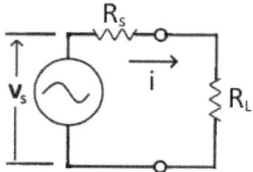

The loop current is,

$$i = \frac{v_s}{R_s + R_L}$$

The power dissipated in $R_L$ is,

$$P_{R_L} = i^2 \cdot R_L = (\frac{v_s}{R_s + R_L})^2 \cdot R_L$$

To find the maximum power delivered to $R_L$, we find the derivative of $P_{R_L}$, set it to zero, and solve for $R_L$.

$$\frac{dP_{R_L}}{dR_L} = v_s^2 \cdot (R_s + R_L)^{-2} \cdot R_L$$

$$= v_s^2 \cdot [-2 \cdot (R_s + R_L)^{-3} \cdot R_L + (R_s + R_L)^{-2} \cdot 1] = 0$$

$$\frac{dP_{R_L}}{dR_L} = v_s^2 \cdot \left[ \frac{-2 \cdot R_L}{(R_s + R_L)^3} + \frac{1}{(R_s + R_L)^2} \right] = v_s^2 \cdot \left[ \frac{-2 \cdot R_L + R_s + R_L}{(R_s + R_L)^3} \right] = 0$$

$$-2 \cdot R_L + R_s + R_L = R_s - R_L = 0$$

$$R_L = R_s$$

We, therefore, conclude that maximum power transfer occurs when the load resistance equals the source output resistance.

# Appendix D – Push-Pull Amplifier Harmonic Distortion

Assume that the input signal to push-pull BJT $Q_1$ is,

$$i_{in1} = \sin(\omega t)$$

and the input to $Q_2$ is 180 degrees out of phase,

$$i_{in2} = \sin(\omega t + \pi)$$

The corresponding outputs are,

$$i_{out1} = I_c + A_0 + A_1\sin(\omega t) + A_2 \sin(2\omega t) + A_3 \sin(3\omega t) + \cdots$$

$$i_{out2} = I_c + A_0 + A_1\sin(\omega t + \pi) + A_2 \sin(2(\omega t + \pi)) + A_3 \sin(3(\omega t + \pi))$$
$$+ \cdots$$

The $I_c$ and $A_0$ terms represent any DC offset and do not contribute to signal output. The "$\omega t$" terms are the desired and amplified original signals with current gain $A_1$. The remaining terms are components of distortion introduced by the non-linearity of the BJTs.

Looking at the $i_{out2}$ term more closely, we see that,

$$i_{out2} = I_c + A_0 + A_1\sin(\omega t + 2\pi) + A_2 \sin(2\omega t + 2\pi)) + A_3 \sin(3\omega t + 3\pi))$$
$$+ \cdots$$

From trigonometry,

$$\sin(n\omega t + n\pi)) = \sin(n\omega t) \text{ if n is even and } -\sin(n\omega t) \text{ if n is odd.}$$

Therefore, we can rewrite the output equations as,

$$i_{out1} = I_c + A_0 + A_1\sin(\omega t) + A_2 \sin(2\omega t) + A_3 \sin(3\omega t) + \cdots$$

$$i_{out2} = I_c + A_0 - A_1\sin(\omega t) + A_2 \sin(2\omega t) - A_3 \sin(3\omega t) + \cdots$$

These outputs combine in the push-pull output transformer to make the outcome $i_{out}$,

$$i_{out1} = i_{out1} - i_{out2} = 2(A_1\sin(\omega t) + A_3 \sin(3\omega t) + \cdots)$$

Thus, the push-pull configuration has eliminated the even harmonics.

# Appendix E – Derivation of Class AB Power Dissipation

The power dissipated by a Class AB amplifier is the same as a Class B amplifier, except that we must add the power dissipated by the quiescent current $I_Q$. Our approach then is to find the power a Class B amplifier dissipates and add the $I_Q$ dissipation.

The first and essential observation is that maximum dissipation in a Class B/AB amplifier does not occur when it delivers maximum output power. Consider the amplifier of Chapter 17. The figure below shows a single cycle of dissipated power at 100% $P_{out}$ and 40% $P_{out}$.

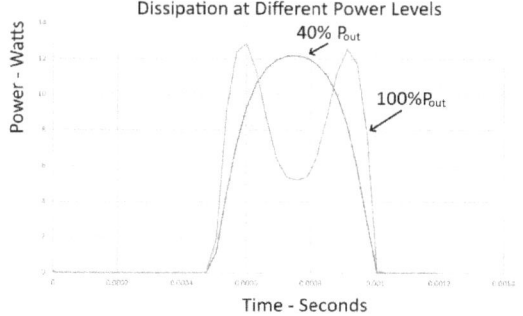

Both curves peak at about 12 watts, but the power curve has a severe dip, as shown. The area under each curve represents the total heat energy delivered to $T_1$'s case. It is this heat energy that must dissipate to prevent unwanted temperature rise.

Because of the dip in the 100% full-power curve, its area is less than the area under the 40% power curve. BJT heating is more significant at 40% maximum power output. This suggests that a point of maximum dissipation must exist between 0% and 100% power output.

To explore this question, we write the general equation for power dissipation in the output BJT $T_1$ of our amplifier.

$$P_D = i_c \cdot v_{CE}$$

Assuming a perfectly balanced BJT pair, the quiescent output voltage is,

$$V_Q = \frac{V_{CC}}{2}$$

We also assume that during the half cycle $T_1$ is conducting, its collector current $i_c$ is,

$$i_c = \frac{V_Q}{R_L}$$

Further, if the driving signal is $\sin(\omega t)$, then the voltage across the collector of $T_1$ is,

$$v_{CE} = V_Q \cdot (1 - \sin(\omega t))$$

This equation states the value of $v_{CE}$ at maximum output. Since we already expect that maximum dissipation does not occur at maximum output, then we introduce a gain factor $0 \le k \le 1$. Then,

$$v_{CE} = V_Q \cdot [1 - k \cdot \sin(\omega t)]$$

When k = 0, the output is also 0. When k = 1, we have maximum output. Later, we maximize the heat produced based on the value of k.

The gain constant k also shows up in the $T_1$'s collector current along with a sinusoidal variation. Thus,

$$i_c = k \cdot \frac{V_Q}{R_L} \cdot \sin(\omega t)$$

$T_1$'s power dissipation then is,

$$P_D = i_c \cdot v_{CE} = V_Q \cdot [1 - k \cdot \sin(\omega t)] \cdot k \cdot \frac{V_Q}{R_L} \cdot \sin(\omega t)$$

Removing the parentheses and regrouping with the common factor, $\frac{V_Q^2}{R_L}$, we find that,

$$P_D = k \cdot \frac{V_Q^2}{R_L} \cdot \sin(\omega t) - \frac{k^2 \cdot V_Q^2}{R_L} \cdot \sin^2(\omega t)$$
$$= \frac{V_Q^2}{R_L} \cdot (k \sin(\omega t) - k^2 \cdot \sin^2(\omega t))$$

As noted above, the flow of heat energy from $T_1$'s junction into the case causes the heating. To find the total heat energy delivered $E_H$, we must find the area under the curve represented by the equation above. This we do by finding the integral of $P_D$.

$$E_H = \frac{V_Q^2}{R_L} \cdot \left\{ k \cdot [-\cos(\pi) - (-\cos(0))] - k^2 \cdot \int_0^\pi \sin^2(\omega t)dt \right\}$$

$$= \frac{V_Q^2}{R_L} \cdot \left( \frac{k}{\pi} - \frac{k^2}{4} \right)$$

Next, we look for a maximum of $E_H$ with respect to k by taking the derivative of $E_H$ with respect to k and setting the result equal to zero.

$$\frac{dE_H}{dk} = \left( \frac{1}{\pi} - \frac{k}{2} \right) = 0 \text{ from which } k = \frac{2}{\pi} = 0.64$$

Therefore, maximum dissipation occurs when the $V_{CE}$ is 64% of $V_Q$. At that point,

$$P_D = \frac{(0.64 \cdot V_Q)^2}{R_L} \approx 40\% \text{ of maximum power output.}$$

Since each output BJT contributes one-half, each dissipation is one-half of $P_D$ or 20% of maximum power output.

Consider the amplifier of Chapter 17 as an example. The peak-to-peak voltage is the supply voltage $V_{cc}$. Using $V_{cc}$ is a worst-case situation as the $v_{outMax}$ in a practical amplifier will never achieve this. Assuming it will, however, the maximum RMS voltage is,

$$v_{rms} = \frac{V_{cc}}{2 \cdot \sqrt{2}} = \frac{30}{2.828} = 10.6 \; vrms$$

$$P_{outMax} = \frac{10.6^2}{8} = 14 \; watts$$

$$P_{D\,T_1} = 0.20 \cdot 14 + 30 \cdot 0.1 = 2.8 + 3.0 = 5.8 \; watts$$

So, we would use 5.8 watts as a maximum dissipation for figuring out heat sink requirements.

# Appendix G – Voltage and Current Divider Laws

Two circuit types found in applications are the *voltage divider* and the *current divider*. The formula associated with each comes in handy when analyzing circuits. We consider the voltage divider first.

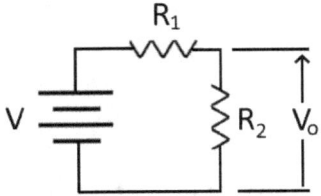

The voltage V divides between the two resistors with $V_o$, the value in which we are interested. To analyze the circuit, we consider the current loop shown below.

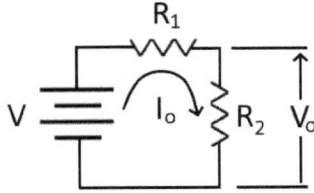

Writing the loop equation, we have,

$$V - (R_1 + R_2) \cdot I_o = 0$$

Solving for $I_o$, we obtain,

$$I_o = \frac{V}{(R_1 + R_2)}$$

Then, $V_o$ is,

$$V_o = \frac{V}{(R_1 + R_2)} \cdot R_2$$

$$V_o = \frac{R_2}{(R_1 + R_2)} \cdot V \quad \text{which is the voltage divider law.}$$

For the current divider analysis, we use this circuit.

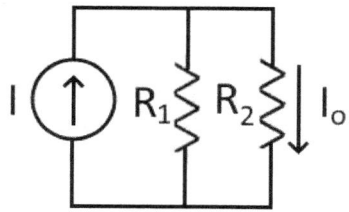

The current I divides between resistors $R_1$ and $R_2$. To analyze, we find the node voltage V below.

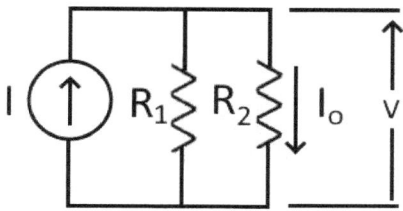

Which is,

$$V = I \cdot \frac{R_1 \cdot R_2}{R_1 + R_2}$$

Current $I_o$ is then,

$$I_o = \frac{V}{R_2} = I \cdot \frac{R_1}{R_1 + R_2} \quad \text{which is the current divider law.}$$

# Appendix H- Thevenin's Theorem

Thevenin's Theorem states that any linear electrical circuit containing only voltage sources, current sources and resistances can be replaced at a pair of terminals by an equivalent combination of a voltage source $V_T$ in a series connection with a resistance $R_T$. The steps for finding $V_T$ and $R_T$ are as follows:

Step 1 – Determine the voltage at the two terminals; call it $V_T$.

Step 2 – Determine the resistance $R_T$ between the two terminals after replacing all voltage sources with a short circuit (or their internal resistances) and all current sources with an open circuit ( or their internal resistances).

Here is an example.

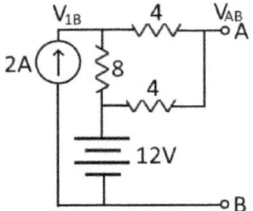

We first find the open circuit voltage $V_{AB}$ by writing a loop equation.

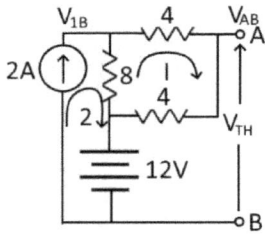

$$(I - 2) \cdot 8 + 4 \cdot I + 4 \cdot I = 0$$

$$8 \cdot I + 8 \cdot I - 16 = 0$$

$$i = 1\,\text{A}$$

Therefore,

$$V_{TH} = V_{AB} = 12 + 4 \cdot 1 = 16\,\text{v}$$

Next, we find $R_{TH}$.

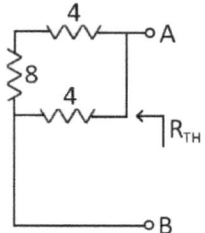

$$R_{TH} = \frac{4 \cdot (8 + 4)}{4 + 8 + 4} = \frac{48}{16} = 3 \, \Omega$$

The Thevenin Equivalent circuit is,

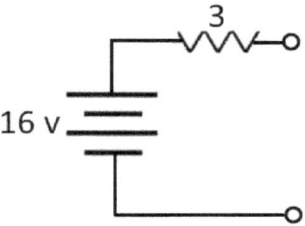

Suppose we wanted to determine the current if we place a 4 $\Omega$ resistor across the output. See the figure below.

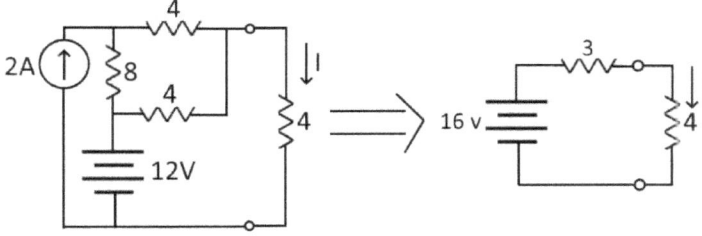

The Thevenin Equivalent makes it easy to solve:

$$I = \frac{16}{7} = 2.286 \, A$$